RAILWAYS OF ARCADIA

1. Lt. Colonel Holman Fred Stephens in his Royal Engineers uniform, probably taken during a Territorial Army Reserve Camp about 1920.

RAILWAYS OF ARCADIA

A Photographic Survey
of the Colonel Stephens Railways
by John Scott-Morgan

with a Foreword by the Series Editor, Dr Paul E Waters

Drawings by Edwin Lambert

and an Appendix on K&ESR Buildings by
Leslie Darbyshire

First Published 1989

ISBN 0 948904 50 X

Designed by D & J Hunter, Beckenham.

Typesetting by Cathy Harper.

Printed in Great Britain by Biddles Ltd, Guildford.

Published by P. E. Waters & Associates, 105, Highland Road, Bromley, Kent, BR1 4AA, England.

1989

Dedication: In memory of Clare Frances Myddleton-Jones to whom I owe so much.

CREDITS AND SOURCES OF PHOTOGRAPHS AND DRAWINGS

J. H. Aston/Lens of Sutton: **76, 180, 193, 264**. W. A. Camwell: **225, 226, 228, 230, 231, 232, 254**. H. C. Casserley: **15, 51, 65, 66, 73, 83, 113, 114, 145, 159, 160, 164, 186, 187, 188, 204, 205, 207, 234, 246, 269**. L. Darbyshire: **89, 90, 273, 274**. C. Hamilton Ellis: **42, 50, 53, 54, 55, 57, 68, 74, 272**. R. G. Jarvis: **61, 161**. John H. Meredith: **92**. A. Mills: **194.** Morey: **26, 27, 28, 29, 33**, H. R. Norman/L&GRP: **82, 87**. F. W. P./Lens of Sutton: **110**. L.W. Perkins: **227**. Tony Riley/SECR Society: **99, 100**. E. E. Smith: **2**. J. L. Smith/Lens of Sutton: **56, 86**. K. A. Stone: **5, 9, 264, 266**. Paul E. Waters: **107, 108, 190, 219, 220, 221, 222, 223, 224, 260, 261, 262, 270**. V. R. Webster: **146**. W. M. West: **254.**

The photographers of the following are unknown:

W. H. Austen collection: **97**. L. Darbyshire collection: **79**. G. Dobell collection: **69**. F. Firth: **239**. I. Gotheridge collection: **34, 72, 155, 169, 185, 229, 253**. Greenwich & District Narrow Gauge (Railway) Society: **256**. E. C. Griffith collection: **3, 25**. P. A. Harding collection: **102**. HMRS: **105, 141**. R. C. Inns collection: **4, 96, 158, 177, 212**. LCGB/Kenn Nunn Collection: **14, 19, 20, 77, 104, 109, 173, 174, 189, 215, 233, 235, 250, 252**. Lens of Sutton: **6, 7, 8, 10, 11, 12, 13, 16, 17, 18, 21, 22, 23, 30, 31, 32, 35, 37, 38, 40, 41, 44, 45, 47, 48, 49, 52, 60, 62, 64, 65, 67, 70, 71, 75, 78, 80, 81, 84, 85, 88, 94, 95, 101, 111, 112, 123, 124, 125, 127, 128, 129, 130, 131, 132, 133, 134 ,135, 136, 137, 138, 140, 142, 143, 144, 147, 148, 149, 152, 153, 162, 163, 165, 166, 167, 170, 175, 178, 179, 181, 184 ,196, 197, 198, 199, 200, 201, 202, 203, 206, 208, 210, 213, 216, 217, 218, 236, 237, 239, 242, 243, 244, 247, 249, 257, 258, 259, 263, 267, 268, 271**. Locomotive & General Railway Photographs: **63, 91, 93, 106, 157, 171, 271**. Locomotive Publishing Company: **58, 59**. Sue McNeilly collection: **191, 195, 214**. John Scott-Morgan collection: **1, 36, 98, 150, 151, 172, 211, 251**. Pamlin Prints: **46**. Photomatic: **24, 36, 39, 154, 156, 176**. Real Photographs: **139, 183, 238, 241, 248**. Stephenson Locomotive Society collection: **168**. H. Vaughan collection: **192, 209**. Welsh Industrial & Maritime Museum collection: **115, 116, 117, 118, 119, 120, 121, 122**.

Unsigned drawings are by Lin Peacock.
Ordance Survey material has been used in the revision of the maps used in this book. Crown copyright reserved.

CONTENTS

FOREWORD

I first came across Col Stephens early in 1966 when a chain of coincidences led me to find a home on the Kent & East Sussex Railway for a preserved BTH Bo-Bo diesel-electric locomotive from the Ford Motor Co Ltd. My early impressions of Stephens, gained from other railway enthusiasts, was of some sort of spiv who ran a number of decrepit, unrenumerative country railways using a collection of totally unsuitable secondhand equipment.

Later, as I began to learn more about his railways and in particular about their extent, I began to realise that the traditional assessment of Stephens was rather far from the mark. To have built up such a large empire of minor railways and to have been the consulting engineer for many more at a time when railway construction had otherwise virtually ceased in this country, indicated that there was more to the man than one might have supposed. The test of a good engineer is the fitness for purpose of his designs, whether they be for a humble industrial shunter or a supersonic aircraft. Stephens's railways were eminently suitable for their purpose and served their local communities well for, give or take a few years, the first quarter of the 20th century. Clearly, then, Stephens meets the criterion for a good engineer. If it were not so, he would not have been consulted on so many railway projects right up to his death in 1931, when he had so recently been working on a light rail extension of the Southern Electric across the North Downs to serve the Biggin Hill area of Kent.

To illustrate the extent of his work, I have prepared an outline chronology and list of Stephens's projects based on his application for transfer to M.I.C.E. and on his printed curriculum vitae of about 1918. It is probable that we will never know the full extent of his activities because of the loss of most of the records from the Tonbridge office.

A fascinating aspect of Stephens is the reason why he came to be involved in this field of minor country railways. Much can now only be conjecture, but the circumstancial evidence sets him in the scene of the Country House Movement and other late flowerings of Victorian Idealism which were destined not to survive beyond the 1930's. Indeed, Stephens only lived to the end of his era and, to all intents and purposes, so did his railways. The remnants that lingered on after him had mostly lost their real purpose to road transport by the mid-1930's.

One of the difficulties now in researching the story of Stephens and his railways is the lack of primary documentary source material. Much of the most useful archive material was destroyed by the staff of 23 Salford Terrace when the last of the active Stephens railways were nationalised. Also few people survived to give a first hand account to modern researchers. Therefore, photographs provide much valuable and irreplaceable evidence. This is my excuse for publishing yet another Colonel Stephens album. If photographs are all we have left to tell the story, why not use them directly?

As I have suggested above, it is only when one takes a comprehensive view of Stephens's work that one fully appreciates his abilities. For this reason, I have asked John to structure the book as a comprehensive survey of the various lines that Stephens built and/or managed. The railways are set out in the chronological order of his involvement with them and, as far as possible, a station by station tour along the routes. This has been a compromise dictated by the availability of suitable photographs and by a natural desire to use previously unpublished photographs as far as possible. Photography was not encouraged on these railways because of the bad publicity which they had received from railway enthusiasts even in Stephens's time. Judging by the number of photographs that have turned up over the years, this was seen by many people as a challenge. However, their quality suggests that most were taken by photographers who hid in the hedge until the coast was clear and then nipped out and took a quick snap!

My intention is to publish a series of three books of which this is the first. So any apparent omissions from this first volume will be made up in the future ones. The second will survey the rolling stock and the third will be as full a history as practical of Stephens and his empire. By then we should have set the record straight and given Stephens his due credit as an engineer.

Dr Paul E. Waters
Bromley, Oct 1987

HOLMAN FRED STEPHENS: AN OUTLINE CHRONOLOGY AND PROJECT LIST

Year(s)		Mileage Projected	Mileage Built	Remarks
1868	Born	n.a.	n.a.	31 Oct, Hammersmith, son of Frederick George Stephens and Rebekah Clara Dacton.
1877-83	University College School	n.a.	n.a.	
1883-84	Private Tutor	n.a.	n.a.	
1884-86	Vitre (France)	n.a.	n.a.	French
1886-87	Carlsruhe (Germany)	n.a.	n.a.	German and Mathematics
1887-88	University College London	n.a.	n.a.	Pupil of Sir A. W. B. Kennedy, Civil Engineering, 2 years.
1888-90	Metropolitan Railway, Neasden	n.a.	n.a.	Pupil J. J. Hanbury, Mechanical Engineering, 2.75 years.
1890-94	Cranbrook & Paddock Wood Railway	–	11.44	Resident Engineer during construction.
1894	Institution of Civil Engineers	n.a.	n.a.	Elected A.M.I.C.E. 4 Dec 1894
1894	Tenterden Railway	31.74	–	In charge of survey and sections.
1895-00	Board of Trade	n.a.	n.a.	Acting Inspector, Railway Department.
1895	Cwm Clyddach & Pantcae-Gurwen Railway	–	–	Proposed. Engineer.
1895-31	Rye & Camber Tramway	2.00	2.00	Engineer and Locomotive Superintendent.
1895-31	Rother Valley Railway (1)	12.63	12.05	Engineer, Locomotive Superintendent & Managing Director.
1896	Gower Light Railway	13.82	–	Passed. Engineer.
1896	Hadlow Railway	11.69	–	Passed. Engineer.
1896-01	Medway Upper Navigation	–	16.00	Maintenance and Reconstruction Engineer.
1897	Central Essex Light Railway	27.64	–	Passed. Engineer.
1897-02	Sheppey Light Railway	8.62	8.62	Engineer.
1897	Selsey Tramway (2)	8.00	8.00	Engineer and Locomotic Superintendent.
1898	Cranbrook & Tenterden Light Railway (1)	21.02	1.56	9.89 miles approved, Engineer.
1898	Kelvedon, Coggeshall & Halstead Railway	9.51	–	2.25 miles approved, Engineer.
1898	St Just, Land's End & Great Western Junction Railway	17.12	–	Rejected, Engineer.
1898	Land's End, St Just & Great Western Junction Railway	22.39	–	Rejected, Engineer.
1898	Hedingham & Long Melford Railway	15.34	–	Rejected, Engineer.
1898	Orpington, Cudham & Tatsfield Light Railway	7.85	–	Passed. Engineer.
1899	Long Melford & Hadleigh Light Railway	15.01	–	Passed. Engineer.
1899	Maidstone & Faversham Junction Light Railway	12.16	–	Passed. Engineer.
c1900	Royal Engineers	n.a.	n.a.	Raised 600 men for South African War.
1901-31	RVR Headcorn Extension (1)	7.89	7.86	Engineer.
1902	Surrey & Sussex Railway	25.70	–	Rejected. Engineer.
1904-10	Bere Alston & Calstock Light Railway (3)	10.22	10.22	Engineer, Locomotive Superintendent and General Manager. (4)
1904	Maidstone & Sittingbourne Railway	11.20	–	Rejected. Engineer.
1906	Middle Temple	n.a.	n.a.	Admitted Student.
1906	Headcorn & Maidstone Junction Railway	10.01	–	Passed. Engineer.
1906	Middle Temple	n.a.	n.a.	Admitted Student.
1906	Headcorn & Maidstone Junction Railway	10.01	–	Passed. Engineer.
1907-31	Shropshire & Montgomeryshire Light Railway	–	24.00	Engineer for reconstruction, Locomotive Superintendent and Managing Director.
1909	Bere Alston & Calstock Railway North Hill Extension	6.89	–	Passed. Engineer.
1909	Burry Port & Gwendraeth Valley Light Railway	–	12.29	Engineer for reconstruction for passenger traffic.

Year(s)		Mileage Projected	Mileage Built	Remarks
1910-25	North Devon & Cornwall Junction Light Railway	19.75	19.75	Engineer.
1911	Isle of Wight Central Railway	–		Engineer during re-organisation.
1911-31	East Kent Light Railways	67.07	18.00	Engineer, Locomotive Superintendent and General Manager.
1911-31	Weston, Clevedon & Portishead Railway	–	14.00	Engineer, Locomotive Superintendent and Managing Director.
1912	Cadbury Light Railway	1.08	–	Engineer for proposed branch of WC&PR.
1913-24	Gower Light Railway	13.68	–	Engineer for propsed revival.
1914	Institution of Civil Engineers	n.a.	n.a.	Transfered to M.I.C.E. 10 March 1914.
1914-16	Royal Engineers	n.a.	n.a.	Commanded 2,400 men and 220 officers. Mentioned in despatches 1916.
1916	Royal Engineers (TR)	n.a.	n.a.	Lieut.-Colonel.
1917-19	Edge Hill Light Railway	11.35	5.72	Engineer.
1917-24	West Somerset Mineral Railway	11.00	–	Director during proposed reconstruction.
1918-31	Ashover Light Railway	6.01	6.01	Engineer.
1918	Shropshire Railways (Shrewsbury & Market Drayton Extension)	22.15	–	Rejected. Engineer.
1918-22	Weston Point Light Railway	0.95	0.95	Engineer.
1922-31	Snailbeach District Railways	–	3.50	Director and Engineer.
1923-31	Festiniog and Welsh Highland Railways.	–	34.47	Civil Engineer and Locomotive Superintendent.
1925-31	Southern Heights Light Railway	15.00	–	Passed. Engineer. (5)
?-?	Association of Railways of Local Interest	n.a.	n.a.	Chairman
1931	Died	n.a.	n.a.	23 Oct at Dover.

Notes:

(1) Completed sections formed the Kent & East Sussex Railway in 1905.
(2) Hundred of Manhood and Selsey Tramway later known as the West Sussex Light Railway.
(3) Plymouth, Devonport and South Western Junction Railway.
(4) Assistant to Galbraith and Church on new construction. Engineer for gauge conversion of East Cornwall Mineral Railway.
(5) Branch of Southern Electric Between Orpington and Sanderstead.

1

Introduction

And Arcadia's just another station
on his twice-a-daily
pere-
grination!

"Farmers' Train" Punch, 3 June 1946

NORTHIAM

The warm evening sunshine glinted over the Rother Valley highlighting the rusty corrugated iron roof of Northiam Station and, across the river, Newenden Village stood bathed in the orange glow of dusk. I walked down the road and passed the rotting remains of what was once a gate on a path leading to the small country station. Walking on past the rusty overgrown run-round loop and up the cracked tarmac slope of the platform end, I peered in through the dust-covered window of the waiting room. Its green and cream painted walls looked clean and well kept in contrast to the dull and tarnished appearance of the outside paintwork. The platform hedge was overgrown and untidy. It was all so very different to the same location thirty years before when a feeling of pride and brave survival against all odds had existed.

Although the Kent & East Sussex Railway had been closed for seven years, it was still possible in 1968 to stand at Northiam or Bodiam (or Shepherdswell on the East Kent Railways) on a warm, sunny summer's afternoon and just listen to the birds and imagine a similar afternoon sixty years before.

On a typical station on a Colonel Stephens line, the platform was neatly kept, with an ashed surface and white-washed brick-faced edge. The building made of corrugated iron or timber, was painted in a mellow shade of buff-brown, its awning supports in a darker shade of brown. Inside the booking hall with its timber-planked walls painted cream, upper part, and dark brown, below, were benches built along two walls in an 'L' shape. Opposite, there was a small ticket window with a hatch on a cord over a small pulley on the ticket office side, where there were racks of Edmonson card tickets of all colours and types. Here in the narrow private sanctum, inhabited only by the station-master, you would find a wall telephone, ticket date punch and a pile of standing orders or goods dockets lying near the ticket rack. Next to the booking hall, in the other direction, was the lamp room and parcels office with a lean-to store bringing up the rear. In most cases the stations only had a gentlemen's urinal; ladies had to wait for the train and hope that it contained a bogie brake third with a lavatory.

At Northiam, along the platform edge, dividing it from the goods yard, ran a neatly cut hedge with its green leaves well pruned and angular. One could stand and look into the distance towards Robertsbridge from the platform with its white-washed edge, down the line curving away to the west past a line-side hut built of old sleepers with an old enamel sign for Foster Clarks Custard nailed to its side.

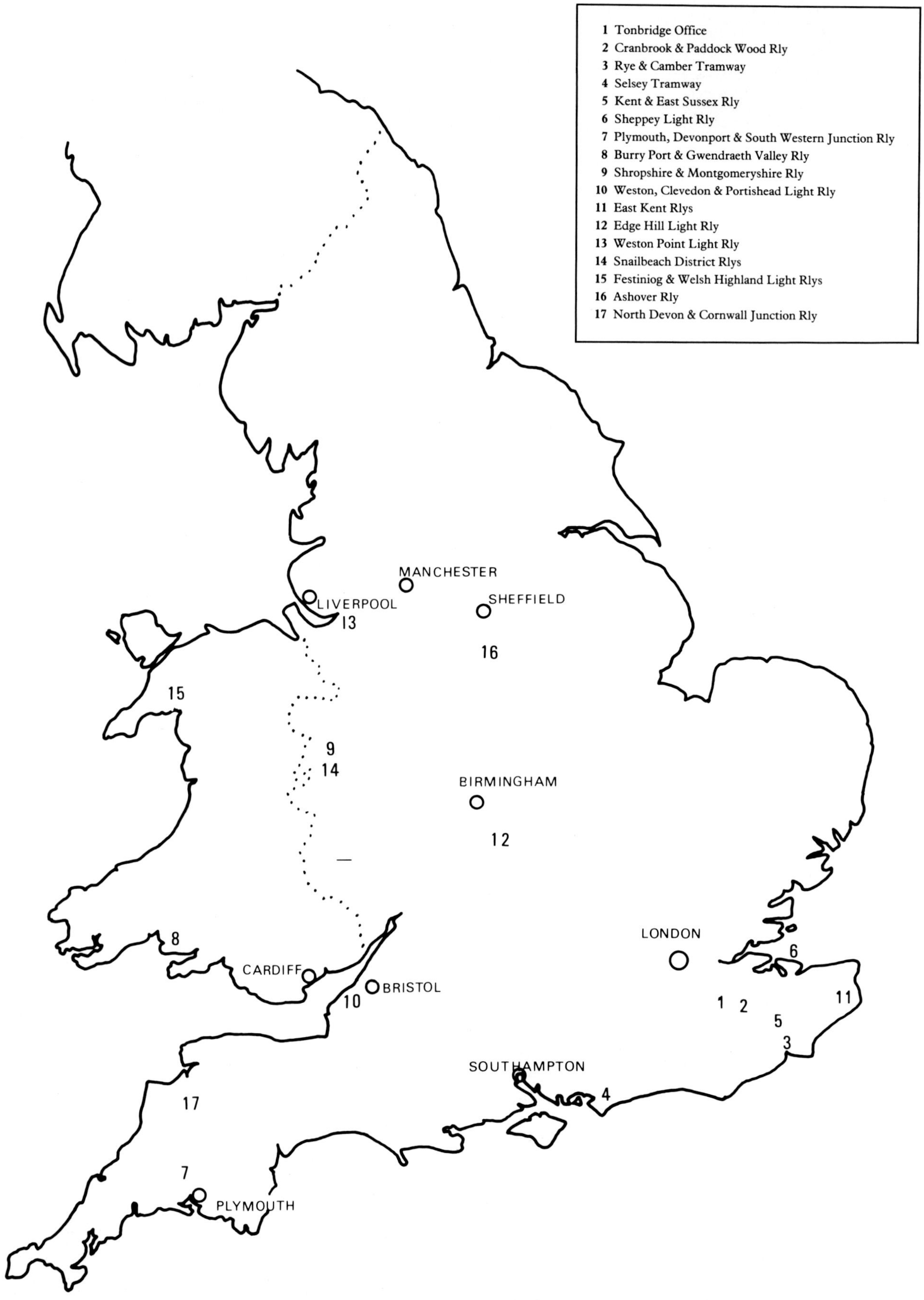

Map showing the locations of Colonel Stephens's railways.

In the goods yard a handful of open wagons stood with their loads of coal being shovelled out by the local merchant, who used the lean-to shed at the end of the building as his office.

It was from here that rural folk came with their local produce and villagers made their occasional trip to London. The farmer, the squire, the local shop-keeper all knew this station and depended on this vital life line that ran across the southern Weald of Kent.

As summer turns to autumn and the leaves on the trees turn golden brown, then also the station takes on a different manifestation. In the dark, early hours of winter and the snows of December, the cheery glow of the fire in the stove, the warm yellow light of the lamps on the platform, that greeted any visitor to these remote parts, brought a basic form of civilisation that seemed to make life tolerable.

The station and the whole line were the product and achievement of a remarkable man who laid the foundations of and built an empire of rural life lines.

HOLMAN FRED STEPHENS

Holman Fred Stephens was born on 31 October 1868 in Hammersmith, West London. His father was Frederick George Stephens, one of the original members of the Pre-Raphaelite Brotherhood, who made his living as art critic.

He grew up in an artistic atmosphere and had a comfortable, if somewhat unconventional, Victorian middle-class upbringing. After his education and training in England, France and Germany, he embarked on an engineering career and built up a group of light railway companies over a period of two decades. The hub of the empire was 23 Salford Terrace, Tonbridge, Kent, an office building a few hundred yards from Tonbridge SER Station. Within its walls the Colonel and his staff managed and operated the group of lines on a daily basis.

Life at Number 23 was full of mixed blessings. The wages were low, below union rates, and there were few perks. Perhaps working for the Colonel provided employment at a time when jobs were very scarce and any salary was better than none. Outside the entrance there was a wooden sign that listed in gilt lettering that this was the headquarters of The Kent & East Sussex Railway, East Kent Railways, West Sussex Railway, Shropshire & Montgomeryshire Railway, Weston Clevedon & Portishead and Festiniog and Welsh Highland Railways.

Stephens had rented these offices from early in 1900 when the practice had become too large to be accommodated at Ashby

2. In the afternoon shadows, O1 0-6-0 No 31370 and a single ex-SE&CR birdcage brake third, bound for Tenterden, trundle into the peaceful station at Biddenden during the last summer (1953) of the existence of the Kent & East Sussex Railway's Headcorn Extension.

3. The festivities at Chichester on the opening day of the Selsey tramway on 27th August 1897.

House. Originally the Colonel rented offices at Ashby House, Prior Road, Tonbridge from 1895, and he still rented rooms there until the time of his death in 1931. He also rented rooms at Robertsbridge, which he rarely used, and was a resident at the Lord Warden Hotel, Dover for many years, which is where he died on 23 October 1931 after a long illness.

A friend of Sir Herbert Walker, Chairman of the Southern Railway, and of Gilbert Szlumper, that Company's Civil Engineer, he was also a friend of Rudyard Kipling, who lived at Batemans, near Etchingham, and to whom he sent an annual free pass to travel on the Kent & East Sussex Railway.

W. H. Austen, his partner, took over the Group after the Colonel's death. Austen had known and worked with Stephens from the early days building the Hawkhurst line.

There are many who think of Stephens as an eccentric almost Colonel Blimp figure. This impression, perpetuated by railway enthusiasts for generations, is not only inaccurate, but unjust. Stephens was, in his way, one of the best railway engineers of the early 20th century. For example, his bridges and buildings were ideal suited to their function. Had his early proposal, for the Rye and Camber Tramway, for a Priestman engined railcar come to pass, he would have been responsible for the first diesel vehicle used on a public railway in Britain.

Stephens specified in detail some very fine steam locomotives which were built by Hawthorn Leslie of Newcastle-

4. On the line to Festiniog, an 0-4-4-0 Fairlie locomotive working hard against the grade with a mixed train c. 1928.

Weston, Clevedon & Portishead Light Railway: Locomotive No. 5 (built Manning, Wardle & Co 1919 works no. 1970) crossing the Triangle, Clevedon, with a train for Weston-Super-Mare, circa 1935.
Edwin Lambert - October 1986

upon-Tyne. For the Kent & East Sussex Railway, he had two attractive 2-4-0T's built, together with perhaps his most beautiful large tank locomotive, the 0-8-0T *Hecate*. *Hecate* was designed to haul heavy freight over the steeply graded section of the line between Rolvenden and Headcorn and ended her days on the Southern. For the Plymouth, Devonport & South Western Junction Railway, he ordered a pair of handsome 0-6-2T's and an 0-6-0T. For the Shropshire and Montgomeryshire Railway he had a pair of smaller well proportioned 0-6-2T'S, both of which later went to the Longmoor Military Railway.

However, his first locomotive, the beautiful 2-4-2T, *Selsey*, was built by Peckett of Bristol, for the Selsey Tramway.

In contrast the Colonel ordered some very basic Ford and Shefflex "tin lizzie" rail buses that rattled and banged their way along three of his lines. Rough riding and smelling of petrol they might have been, but they worked and cut the cost of operation. The Weston Clevedon & Portishead Railway had two purpose built Drewry railcars, one of which (bought after Stephens's time) was the former experimental Southern Railway vehicle, once used on a number of branch lines in Kent, before going to North Somerset.

The Colonel's railways are history now, apart from a few parts of the Kent & East Sussex, East Kent and Plymouth, Devonport and South Western Junction Railways. Modern day railway preservation is fickle and plastic and no substitute for the real thing. It is from archives and old photographs that the truth comes to light and it is from these that we derive the true atmosphere and meaning of railway history. I hope that the reader will find a little of that truth here.

I should like to thank the following people and organisations for their kind help in producing this volume:

Lens of Sutton, LCGB Ken Nunn and Hopwood Collections, LPC/Ian Allan, L&GRP, The Historical Model Railway Society, The Welsh Industrial & Maritime Museum, Ivor Gotheridge, Robert Inns, Richard Stumpf, Jack Slynn, J. Jarvis, Barry A. F. Clark, Tony Riley, W. A. Camwell, P. A. Harding, Sue McNeilly and not forgetting Julia Jacobus who typed the original manuscript.

Special thanks go to Leslie Darbyshire, whose help included the excellent set of drawings of Kent & East Sussex Railway buildings which were drawn or redrawn specially for this book, Edwin Lambert and Lin Peacock, who drew the maps and track plans in the main text, and Dr Paul E. Waters, without whose help in identifying the locations of many of the Plymouth Devonport & South Western Junction, Shropshire & Montgomeryshire and East Kent Railways photographs, in researching and drawing several of the maps and track plans, providing the chapter on the Weston Point Light Railway, in advising on the general structure of the book, in "ghosting" the text and in generally trying to eliminate the mistakes, this book would never have been published.

Finally, I should like to pay tribute to those people who sadly are no longer with us, who also helped with this book, Cuthbert Hamilton Ellis and the Rev. E. R. (Teddy) Boston, among them, gone, perhaps, but not forgotten.

John Scott-Morgan Woking, Sept 1987

2

The Cranbrook and Paddock Wood Railway

The Cranbrook and Paddock Wood Railway was opened from Paddock Wood to Hope Mill (Goudhurst) 1 October 1892 and to Hawkhurst on 4 September 1893. It was a standard gauge line 11 miles 3 furlongs and 5 chains long and it was worked by the SE&CR who absorbed it in 1900. It was finally closed on 10th June 1961. Stephens was resident engineer during the construction of the line from 1890 to 1894 and this was his first professional appointment.

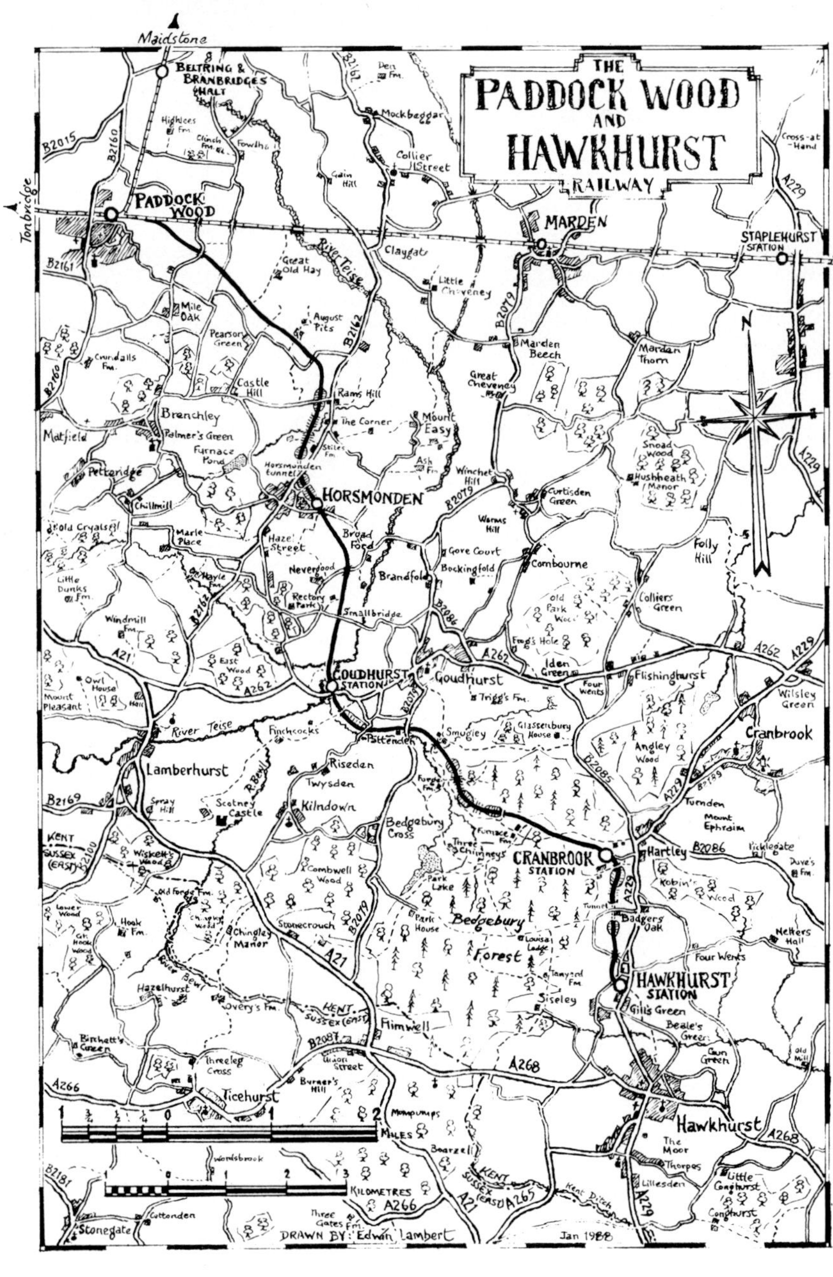

5. Spring in the hopfields: 0-4-4T no. 31517 on a train of bogie stock near Paddock Wood, c. 1955.

6. Horsmonden Station in 1960 showing the Stephens style of building, built by a local builder, Manctelow Bros. of Horsmonden. The barge-boarding on the gabled roof is of interest as being of the same design as that used on the buildings on the Rye & Camber Tramway.

7. Goudhurst Station with its substantial brick SER style station house, but with the booking hall and offices in Stephens/Manctelow Bros style. This station was originally called Hope Mill when opened in 1892.

8. Cranbrook Station looking towards Hawkhurst, c. 1960.

9. Not long before the closure an H Class 0-4-4T and motor set at Cranbrook with a train for Hawkhurst, c. 1960.

10. Cranbrook from the overbridge at the south end of the station during SECR days, c. 1918.

11. The terminus of the Hawkhurst Branch which was a good mile from the town. An ex-LC&DR R1 Class 0-4-4T and ex-SE&CR converted auto set stand waiting for any potential passengers. A bogie composite 3rd, ex-SE&CR stands in the bay platform. Towards the mid-1950's the R1's were replaced by H Class 0-4-4T's.

12. Hawkhurst in the 1920's.

13. A pre-grouping photograph showing the approach to Hawkhurst station.

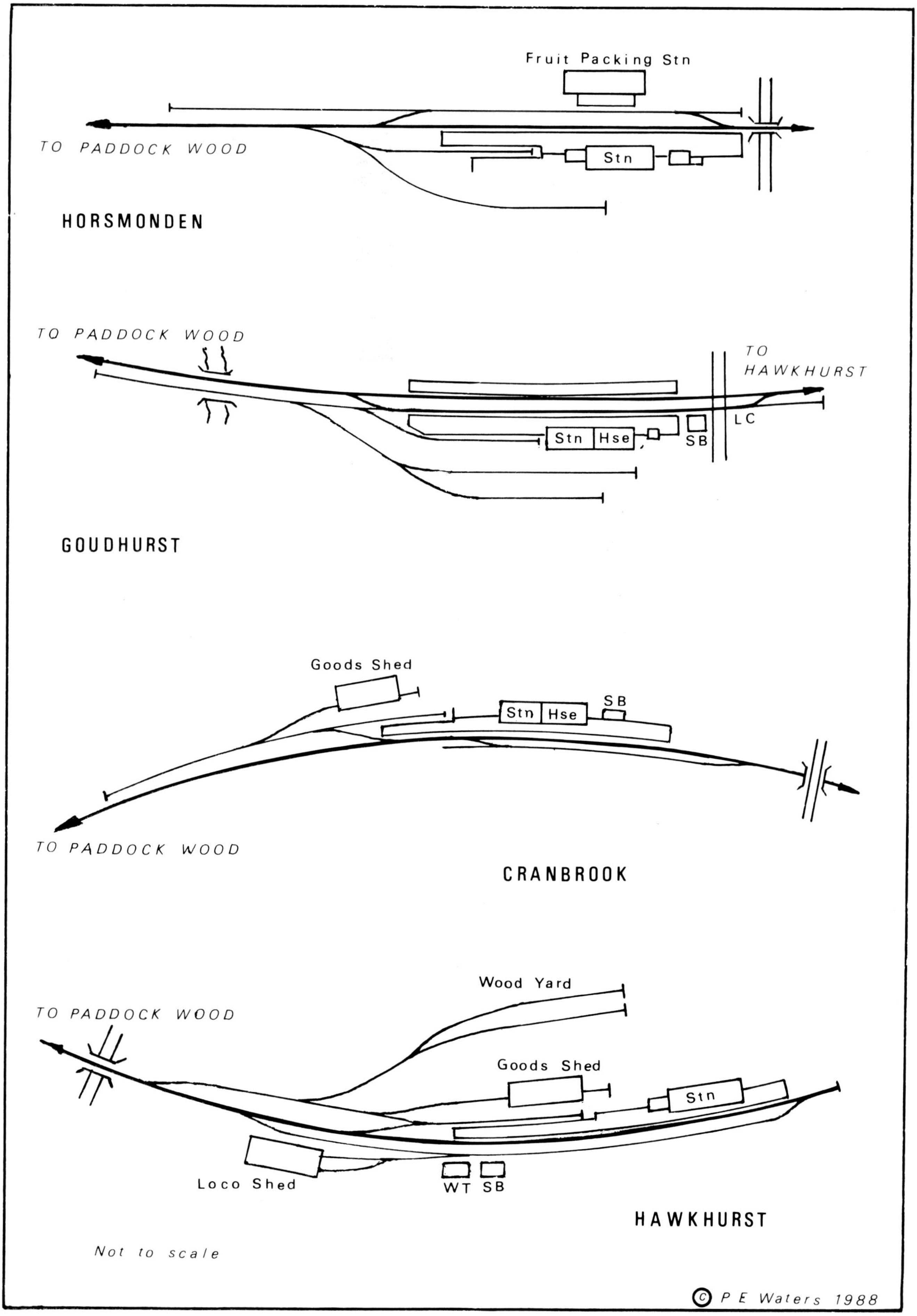
Fruit Packing Stn
TO PADDOCK WOOD
Stn
HORSMONDEN
TO PADDOCK WOOD
TO HAWKHURST
Stn
Hse
SB
LC
GOUDHURST
Goods Shed
Stn
Hse
SB
TO PADDOCK WOOD
CRANBROOK
Wood Yard
TO PADDOCK WOOD
Goods Shed
Stn
Loco Shed
WT
SB
HAWKHURST
Not to scale
© P E Waters 1988

3

The Rye and Camber Tramway

The Rye and Camber Tramway was a 2.5 mile long, 3 foot gauge line opened from Rye to Golf Links on 13 July 1895 and to Camber Sands on 13 July 1908. It closed on 4 September 1939 and was taken over by the military. Stephens was the civil engineer responsible for construction of the line and Manktelow Bros were the contractors. Later Stephens also became the Locomotive Superintendent.

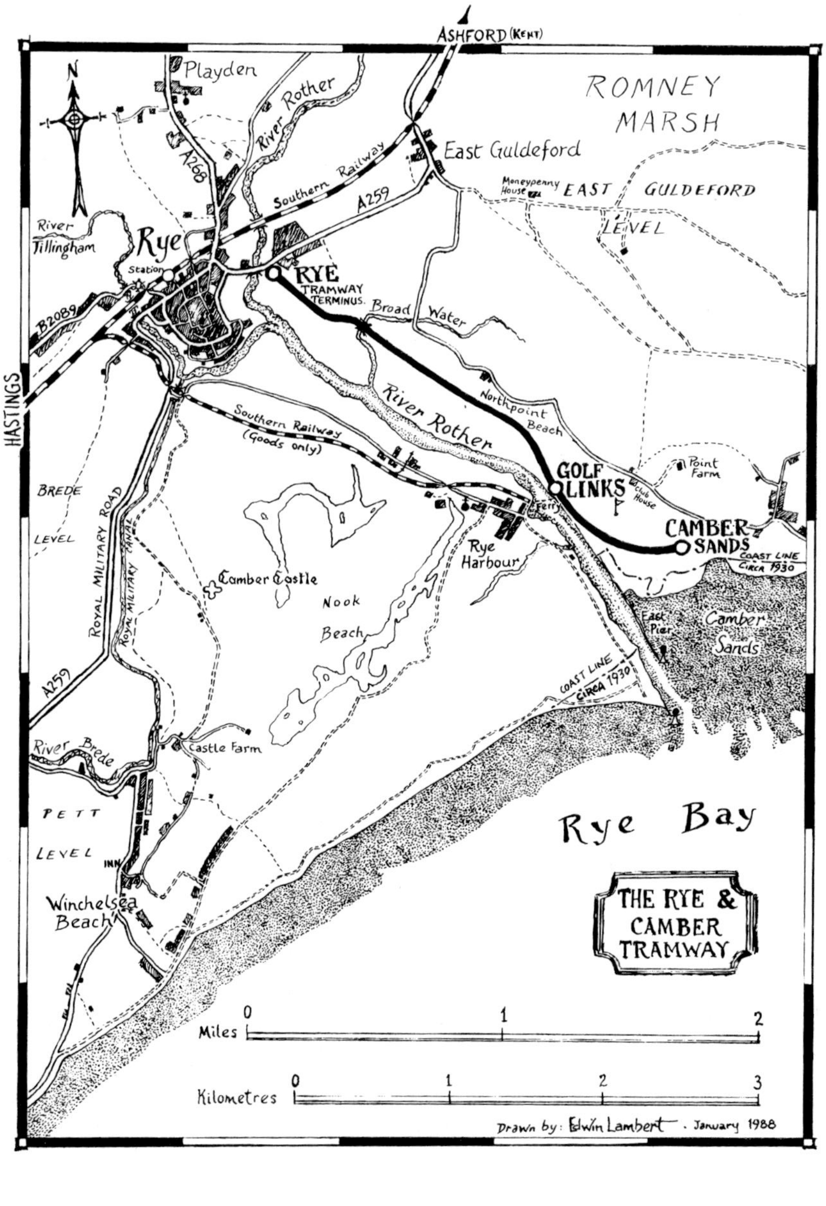

14. Bagnall 2-4-0T, *Victoria*, and both carriages well filled with passengers at Rye Station about to depart for Camber c. 1920.

15. The petrol tractor with both carriages at Rye on 11 April 1931. The carriage next to the locomotive was built by Bagnalls and the other by the Rother Iron Works.

16. The petrol locomotive and both carriages at Rye station c. 1936.

17. An early photograph of 2-4-0T, *Camber*, still in original lined apple green livery and with its original slender, tapered chimney at Rye Station. Note the full stop after its name.

18. After the war, the military handed the Tramway back to the Company. However, the line was in such poor condition that the Company decided not to resume services. Here we see the petrol tractor, which was supplied by the Kent Construction Company in 1925, and both carriages c. 1946.

19. 2-4-0T, *Victoria*, and both carriages crossing Broadwater Bridge heading towards Rye on 18th July 1914. Note the construction of the bridge from I-section beams supported on simple concrete abutments. This was typical of Stephens's low cost bridges.

20. Golf Links Station about 1920.

21. Camber Sands Station, described on Rye & Camber Tramway timetables as being far from the madding crowd. The hut behind the platform used to sell tea and sticky currant buns in the summer.

4

The Selsey Tramway

The Hundred of Manhood and Selsey Tramway was opened from Chichester to Selsey, 7.75 miles, on 27 August 1897 and later extended 1.25 miles to Selsey Beach. This standard gauge line changed its name to the West Sussex Railway in 1924, but was always known simply as the Selsey Tramway. It closed on 19 January 1935. Stephens was the line's Engineer and Locomotive Superintendent from its inception. He later acquired a controlling interest in the Company and was Chairman by the 1920's.

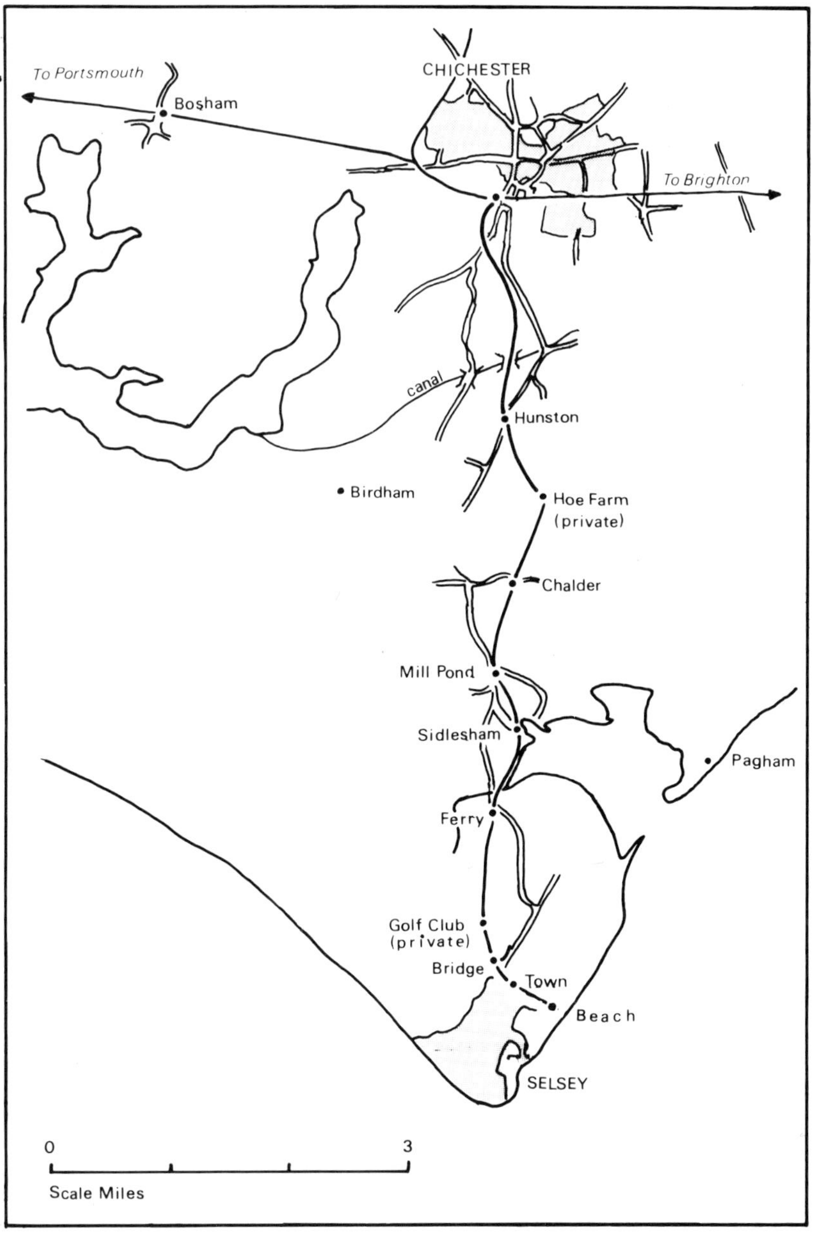

22. 2-4-2T No. 1, *Selsey*, and train of Falcon bogie stock at Chichester c. 1910.

23. *Sidlesham* at Chichester with a train containing two ex-Lambourn Valley Rly. 4-wheel coaches c. 1915.

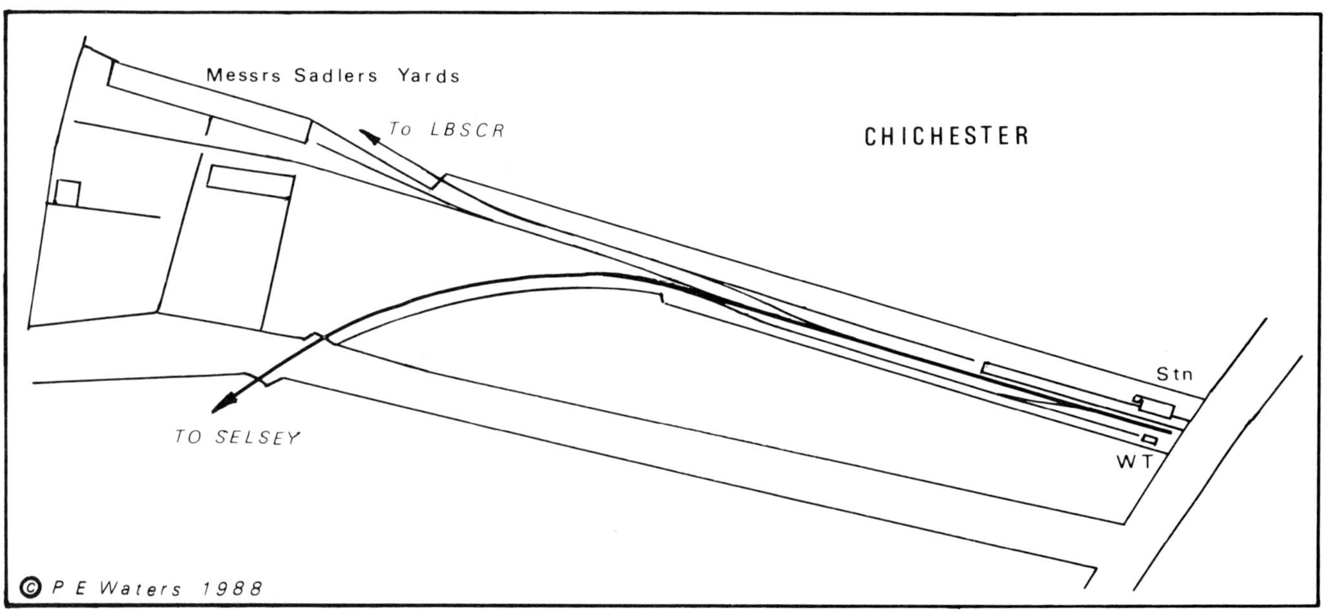

24. Chichester Station, with its well known gas holders, derelict in the spring of 1935.

25. The lifting bridge across the Chichester Canal, with a coastal vessel about to pass through, c. 1897

26. Sidlesham Station during the great flood, 16 December 1910.

27. The new embankment at Sidlesham constructed to raise the line above the level of the floods.

28. A view of the accident near Golf Club Halt on 3 September 1923. This clearly shows the damaged Hudswell 0-6-0ST, *Chichester*, with its bunker pushed foward by the buffer of the leading ex-LC&DR 4-wheeled carriage, which pinned the fireman to boiler backhead.

29. A second view of the derailment on the 3 September 1935, showing how the flat bottom rail had been spread by the train. Note the very poor condition of the track with rotting sleepers. The driver escaped with scalds, but the fireman was killed.

30. Over the weed choked track runs the Shefflex set on its way from Selsey to Chichester with a luggage trolley full of local goods c. 1930.

31. Far from the madding crowd, a petrol railcar set made up of a Shefflex car leading a Ford with a luggage trolley between trundles through the trees near Selsey c. 1934.

32. Selsey Station shortly after the line opened showing the original round roofed locomotive shed.

33. 0-6-0ST, *Sidlesham*, and mixed train at Selsey c. 1910.

34. General view of Selsey station c. 1925.

35. Selsey Station just before the Great War with 0-4-2ST No. 3 and train of bogie stock. Note the enlarged station building with 2 windows in the gabled end.

36. Selsey Station shortly after the line's closure to traffic, showing the station building and the small goods shed, spring 1935.

37. 0-6-0ST, *Morous*, at Selsey with a mixed train consisting of a private owner wagon and an ex-LC&DR coach, c. 1930.

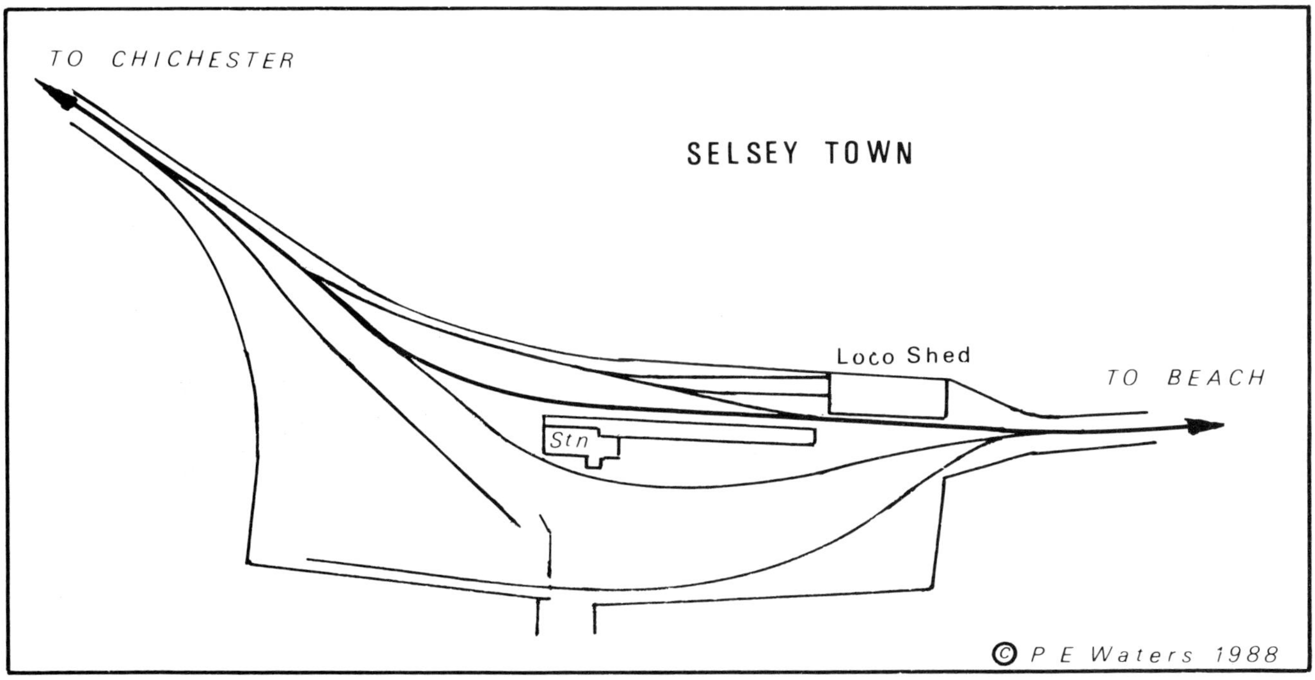

38. *Sidlesham* receives attention at Selsey shed c. 1925. Note the "Private" notice nailed to the main upright.

39. A dilapidated Selsey shed in the spring of 1935.

40. Three from Leeds at Selsey shed c. 1934: Manning Wardles Nos. 5, *Ringing Rock*, and 2, *Sidlesham*, with Hudswell Clarke, No. 4, *Chichester*, between.

5

The Kent and East Sussex Railway

The Rother Valley Railway, later the Kent & East Sussex Railway was a standard gauge line 21 miles long. The Rother Valley Railway opened to Rolvenden on 2 April 1900, from Rolvenden to Tenterden 15 April 1903 and from Tenterden to Headcorn on 15 May 1905 when the name was changed. The railway closed from Headcorn to Tenterden on 2 January 1954, when the passenger service was withdrawn from the rest of the line, and totally on 22 April 1961. Bodiam to Tenterden is now preserved. Stephens was Engineer, Locomotive Superintendent and later Managing Director.

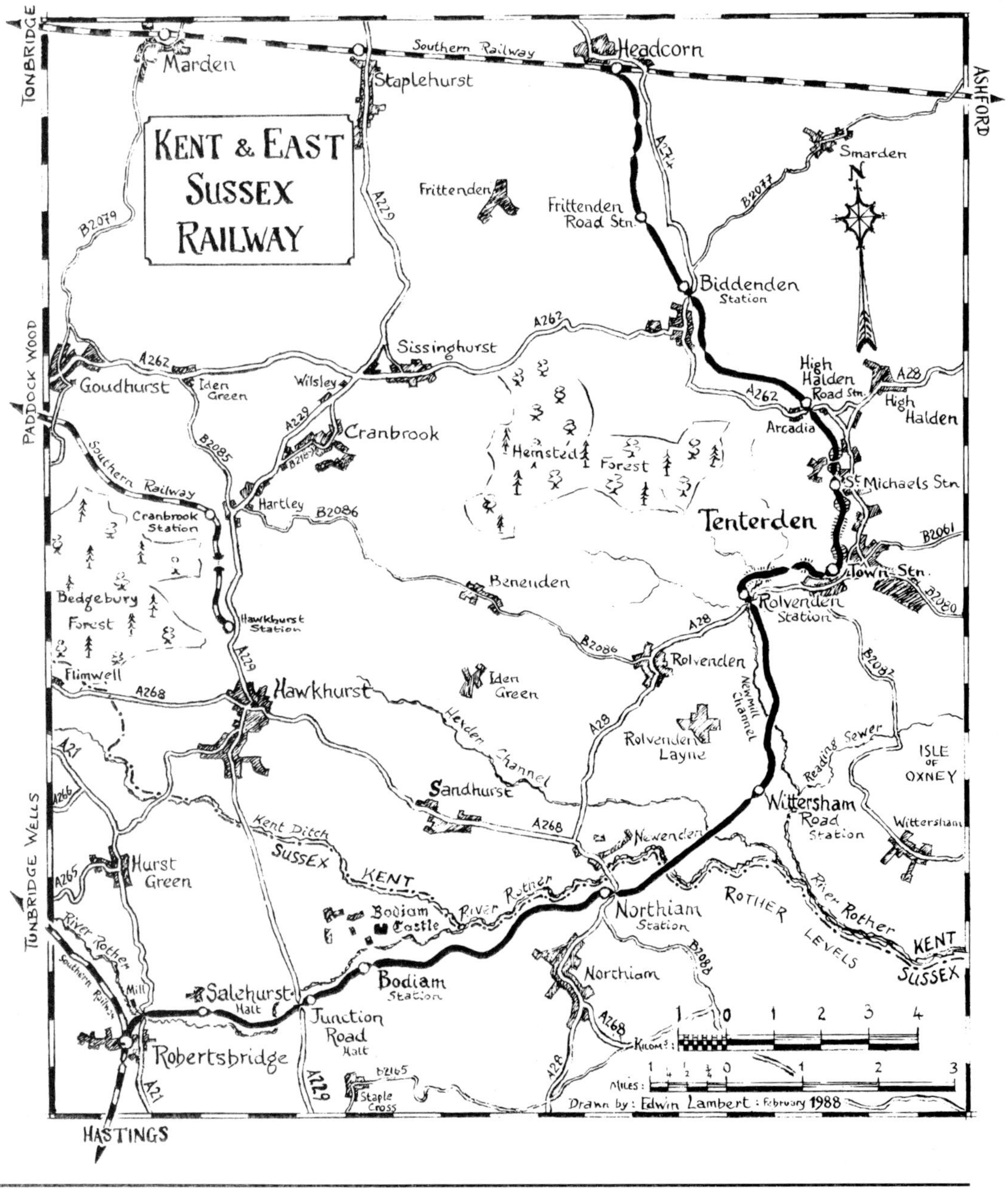

41. On a warm summer afternoon in Aug 1950, A1X Terrier 0-6-0T No. 32678 stands in the bay platform at Robertsbridge with an ex-LSWR bogie corridor brake-3rd awaiting the road along the Rother Valley to Tenterden Town. Note that the Terrier is in full lined out BR livery without its BR insignia. 32678 was on the Isle of Wight from the early 1920's, hence its enlarged coal bunker.

42. Ex-SE&CR P Class 0-6-0T No. 1325 on loan from the SR in the up main line platform at Robertsbridge while engaged in shunting c. 1946. The K&ESR had running rights over the main line to gain access to the industrial siding just south of the level crossing.

43. 0-6-0T No. 3, ex-LBSCR A1 Class, arrives at Robertsbridge with an ex-LSWR bogie brake 3rd c. 1935.

44. Terrier No. 32678 and mixed train climbing the short incline on its approach to Robertsbridge c. 1949.

45. No. 32678 again shunting at Hodsons Mill, Northbridge Sreet, Robertsbridge. This mill had the only industrial siding on the K&ESR. After closure of the K&ESR in 1961, the mill provided its own motive power, an ex-SE&CR P Class 0-6-0T No. 1556, which was by chance the first locomotive hired by the K&ESR in the early 1930's, and worked the line to Robertsbridge Station as part of its industrial siding.

46. Solitude at Salehurst Halt, July 1958. This lonely sleeper built platform served a pub, church and small cluster of houses about 1.5 miles from Robertsbridge.

47. The original sleeper and ash platform at Junction Road Halt c 1920.

48. The later concrete platform at Junction Road. Showing similarity to Southern Railway platforms, it is believed to have been built (by BR in 1948) mainly from parts made at Rolvenden in 1946.

49. A peaceful rural scene at Bodiam in the mid-1920's with the goods yard full of coal wagons and hop fields in the background. Note the enamel advertisements for Mazawatee Tea and Suttons Seeds, reflecting the sun, and the dark blue station name board with white lettering. There was no loop here; the line in the foreground was a siding facing Tenterden.

50. Long ago and far away a group of hop pickers pose by engine No. 4 and its train of two ex-L.S.W.R. bogie non-corridor thirds at Bodiam.

51. On 14 March 1931 a Robertsbridge-bound Ford railbus set approaching Bodiam from Northiam past the home signal.

52. A winter's day at Northiam c. 1925, showing the coal yard full of wagons. The second platform, on the loop, was later removed and a hedge replaced the fence on the main platform. Northiam Station is nearer to Newenden (1/4 mile) than Northiam (1.5 miles).

53. Terrier No. 3, *Bodiam*, and mixed train at Northiam c 1936. The locomotive blows off in the warm afternoon sunshine while waiting to leave for Tenterden.

54. On another warm afternoon in July 1936 at Northiam, No. 3, with is copper capped chimney burnished, if a little sooty, bound for Robertsbridge with an ex-L.S.W.R. bogie brake third, simmers under the heat of the sun, while the station-master walks up the platform with the single line staff.

55. On a winter's morning in 1934, No. 3 looms out of the mist lying low over the Rother Levels on its approach to Northiam with a mixed train for Robertsbridge. This is a good photograph of the Northiam home signal, which used to stand on the Wittersham side of the level crossing.

56. Wittersham Road was meant to serve the village of Wittersham in the Isle of Oxney, some 1.5 miles away. The station building was at right angles to the platform and faced the road. Behind the platform was a goods yard used in the Second World War by a rail mounted gun (1940-42).

57. The double armed, slotted post signal at Wittersham Road, c. 1936.

58. Rother Valley Railway 2-4-0T No. 1, *Tenterden*, and train of new 4-wheeled stock, brake van No. 7, third No. 2 and first No. 5 leading with third No. 4 and brake van No. 8 at the rear, at Tenterden Station (as Rolvenden was then called) at about the time of the opening of the line in April 1900. Note that the RVR brake van has a garter crest and a different style of lettering to that used by the builders in their works photographs.

59. The original Rolvenden Station showing the round roofed locomotive shed, of virtually the same design as that at Selsey, containing locomotive No. 2. The brake van, No. 8, and third saloon coach, No. 4, appear to be the end of the same posed train as in photograph 58 suggesting that it was taken on the same day in c. 1900. Note the close proximity of the Newmill Channel, which was later diverted to dog-leg round a much enlarged area to allow for carriage sidings and a paint shop.

60. A panorama of Rolvenden yard c. 1935, showing the dump with the Pickering steam railcar just visible behind the cab of an Ilfracombe Goods on the left and Shefflex railbus outside the carriage shed, left of centre.

61. No. 3 shunts a mixed train containing an ex-LSWR Bk, 3rd at Rolvenden. This was the 16.30 to Robertsbridge on 27 March 1937.

62. 0-6-0ST No. 8, *Hesperus*, formerly *Ringing Rock*, running as an 0-4-2ST entering Rolvenden yard from Robertsbridge with a mixed train containing ex-Great Eastern 4-wheelers c. 1928.

63. Rolvenden shed c. 1935 with ex-LSWR 0-6-0ST No. 4 peeping out of the doors with smoke drifting from its tall chimney. Note the stored boilers from the Ilfracombe Goods 0-6-0's and that the shed now has a pitched roof.

64. Under the shadows a stored No. 2, *Northiam*, stands c. 1930. Behind her No. 1, *Tenterden*, undergoes a heavy overhaul.

65. From the mid-1930's the Company had to hire locomotives from the Southern. Here we see P Class 0-6-0T No. 1325 with No. 4, the ex-LSWR 0-6-0ST, at Rolvenden in c. 1946.

66. Ex-K&ESR No. 3 as BR No. 32670 at Rolvenden with a mixed train for Robertsbridge on 15 July 1950. Note the ex-boat train all third in the siding behind.

67. An afternoon mixed train, headed by O1 Class 0-6-0 No. 31048, starts its journey from Rolvenden to Headcorn in the winter of 1953, a few months before the end.

68. No. 4 and No. 3 with an ex-LSWR non-corridor bogie brake between ascend Tenterden Bank c. 1934. This formation was often used on the K&ESR when two locomotives had to work light to spread the load on the bridges, especially on the RVR Section where the bridges were designed for a 10 ton axle load.

69. A rare photograph of the original corrugated iron station building at Tenterden Town in the winter of 1902/3 during the building of the extension from Rolvenden. This structure was later replaced by the familiar brick building and re-used at Headcorn.

70. An early view of Tenterden Town Station (c. 1910 ?) showing the later brick building, containing a booking hall in the centre, booking office to the right and lamp and parcels room to the left, and the three arm signal.

71. Another view of Tenterden Town station showing the three arm signal, second platform and wind-driven water pump c. 1930. The crossing gates were later replaced by cattle grids.

72. A general view of Tenterden Town Station c. 1920 showing all of the platforms.

73. On loan from the Southern, 0395 class 0-6-0 No. 3440 at Tenterden Town awaiting the road to Robertsbridge on 26 April 1947.

74. No. 2 with a single carriage train waits in the afternoon sun at Tenterden Town for the road to Robertsbridge c. 1936.

75. St Michaels looking towards the tunnel. This Halt served a northern suburb of Tenterden.

76. Rustic charm at Tenterden St Michaels Halt in August 1953.

77. The short tunnel at St Michaels looking towards Headcorn Junction. Note the telephone insulators on the wall.

78. High Halden Road Station in the 1920's showing the slotted post signal, neat gardens, milk churns on the platform and the timber station building.

79. 0-6-0ST No. 8 and mixed train at High Halden Road c. 1930. The leading vehicle is an ex-LSWR non-corridor brake third.

80. High Halden Road in November 1953, just before closure, with 01 Class 0-6-0 No. 31065 on a mixed train for Tenderden Town. Note the closure notice on the fence.

81. Biddenden Station c. 1908 with one of the original Rother Valley Railway 2-4-0T's and a set of 4-wheeled stock which includes two ex-NLR brake vans and ex-MS&LR vehicles. Biddenden had a loop (foreground) and a goods yard behind the station.

82. Biddenden Station looking towards Headcorn 15 Aug 1931, showing the passing loop and the second platform. Note the mixture of lamps, oil for the platform and gas for the station approach.

83. Biddenden from the train on 15 July 1950.

84. Frittenden Road Station facing towards Headcorn c. 1930. This lonely station served a village about 2 miles away. Note the unusual two arm signal and the timber faced platform. It was to the station-master here that Stephens wrote on 3rd October 1908, after complaints from the SE&CR at Broadstairs in connection with local parcels error notes, ". . . You have next door to nothing to do at your station and the least that you can do is to do the little you have to do properly. Please take this as a warning."

85. Ex-LSWR 0-6-0ST No. 4 at Headcorn in the exchange sidings on a freight transfer c. 1934.

86. O1 0-6-0 No. 1064 with a single ex-LSWR bogie brake 3rd at Headcorn on 3 July 1947.

87. The newly rebuilt Headcorn Junction Station on 15 August 1931 showing the corrugated iron building from Tenterden Town and the Southern Railway concrete fence posts and platform edging.

88. Another view of the newly rebuilt station at Headcorn in the 1930's.

89. The original RVR signal at Bodiam facing Robertsbridge, 1955.

90. Southern rail-built home signal near Salehurst, 1955.

91. The slotted post signal at Wittersham Road, 1929.

92. The three arm signal at Tenterden Town, c. 1953.

93. Shunt signal at Rolvenden, c. 1946.

94. A Kent & East Sussex Austen period cast trespass notice dated March 1946.

95. A Rother Valley Railway enamel trespass notice.

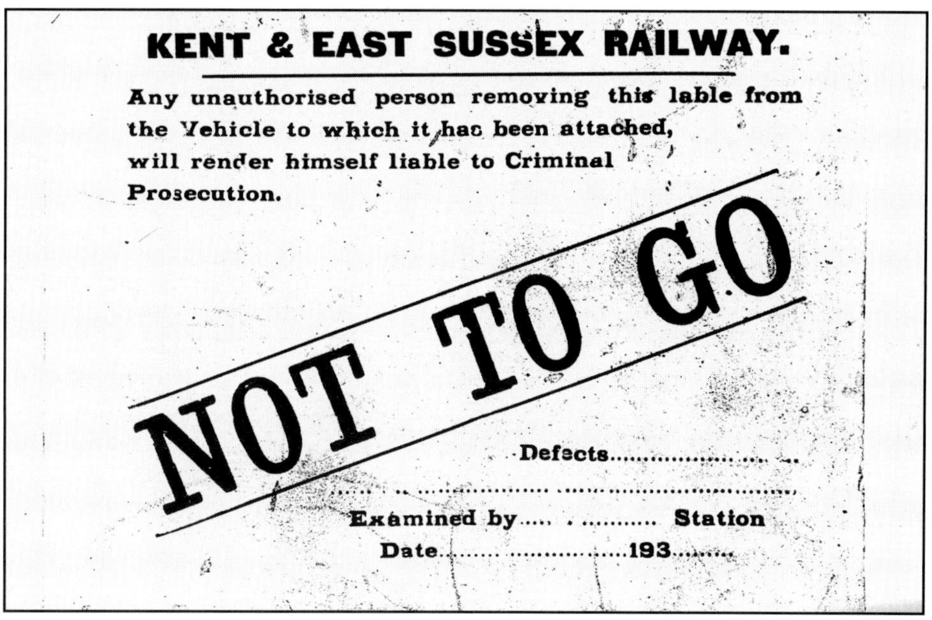

KENT & EAST SUSSEX RAILWAY.

Any unauthorised person removing this lable from the Vehicle to which it has been attached, will render himself liable to Criminal Prosecution.

NOT TO GO

Defects........................

..

Examined by................ Station

Date...........................193.........

96. A Kent & East Sussex Railway Wagon card, c. 1930.

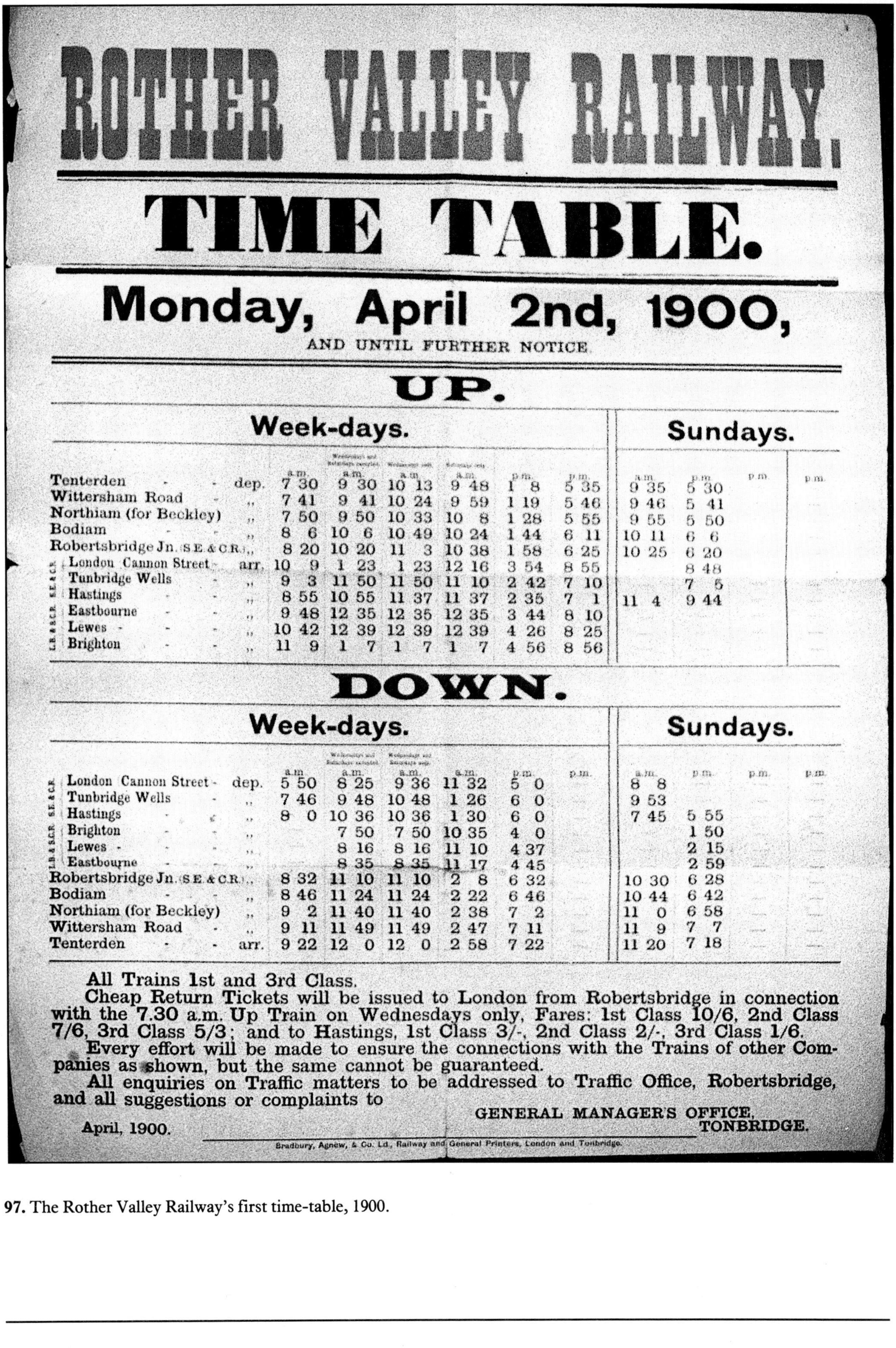

ROTHER VALLEY RAILWAY.

TIME TABLE.

Monday, April 2nd, 1900,

AND UNTIL FURTHER NOTICE.

UP.

		Week-days.						Sundays.			
		a.m.	a.m.	a.m.	a.m.	p.m.	p.m.	a.m.	p.m.	p.m.	p.m.
Tenterden	dep.	7 30	9 30	10 13	9 48	1 8	5 35	9 35	5 30		
Wittersham Road	,,	7 41	9 41	10 24	9 59	1 19	5 46	9 46	5 41		
Northiam (for Beckley)	,,	7 50	9 50	10 33	10 8	1 28	5 55	9 55	5 50		
Bodiam	,,	8 6	10 6	10 49	10 24	1 44	6 11	10 11	6 6		
Robertsbridge Jn. (S.E. & C.R.)	,,	8 20	10 20	11 3	10 38	1 58	6 25	10 25	6 20		
S.E. & C.R. London Cannon Street	arr.	10 9	1 23	1 23	12 16	3 54	8 55		8 48		
S.E. & C.R. Tunbridge Wells	,,	9 3	11 50	11 50	11 10	2 42	7 10		7 5		
S.E. & C.R. Hastings	,,	8 55	10 55	11 37	11 37	2 35	7 1	11 4	9 44		
L.B. & S.C.R. Eastbourne	,,	9 48	12 35	12 35	12 35	3 44	8 10				
L.B. & S.C.R. Lewes	,,	10 42	12 39	12 39	12 39	4 26	8 25				
L.B. & S.C.R. Brighton	,,	11 9	1 7	1 7	1 7	4 56	8 56				

DOWN.

		Week-days.						Sundays.			
		a.m.	a.m.	a.m.	a.m.	p.m.	p.m.	a.m.	p.m.	p.m.	p.m.
S.E. & C.R. London Cannon Street	dep.	5 50	8 25	9 36	11 32	5 0		8 8			
S.E. & C.R. Tunbridge Wells	,,	7 46	9 48	10 48	1 26	6 0		9 53			
S.E. & C.R. Hastings	,,	8 0	10 36	10 36	1 30	6 0		7 45	5 55		
L.B. & S.C.R. Brighton	,,		7 50	7 50	10 35	4 0			1 50		
L.B. & S.C.R. Lewes	,,		8 16	8 16	11 10	4 37			2 15		
L.B. & S.C.R. Eastbourne	,,		8 35	8 35	11 17	4 45			2 59		
Robertsbridge Jn. (S.E. & C.R.)	,,	8 32	11 10	11 10	2 8	6 32		10 30	6 28		
Bodiam	,,	8 46	11 24	11 24	2 22	6 46		10 44	6 42		
Northiam (for Beckley)	,,	9 2	11 40	11 40	2 38	7 2		11 0	6 58		
Wittersham Road	,,	9 11	11 49	11 49	2 47	7 11		11 9	7 7		
Tenterden	arr.	9 22	12 0	12 0	2 58	7 22		11 20	7 18		

All Trains 1st and 3rd Class.

Cheap Return Tickets will be issued to London from Robertsbridge in connection with the 7.30 a.m. Up Train on Wednesdays only, Fares: 1st Class 10/6, 2nd Class 7/6, 3rd Class 5/3; and to Hastings, 1st Class 3/-, 2nd Class 2/-, 3rd Class 1/6.

Every effort will be made to ensure the connections with the Trains of other Companies as shown, but the same cannot be guaranteed.

All enquiries on Traffic matters to be addressed to Traffic Office, Robertsbridge, and all suggestions or complaints to

GENERAL MANAGER'S OFFICE,
TONBRIDGE.

April, 1900.

Bradbury, Agnew, & Co. Ld., Railway and General Printers, London and Tonbridge.

97. The Rother Valley Railway's first time-table, 1900.

6

The Sheppey Light Railway

The Sheppey Light Railway was a standard gauge line, 8.75 miles long from Queenborough to Leysdown, opened on 1 August 1901. Stephens was the engineer, but after the opening had no further connection with the line which was worked by and eventually absorbed by the SER.

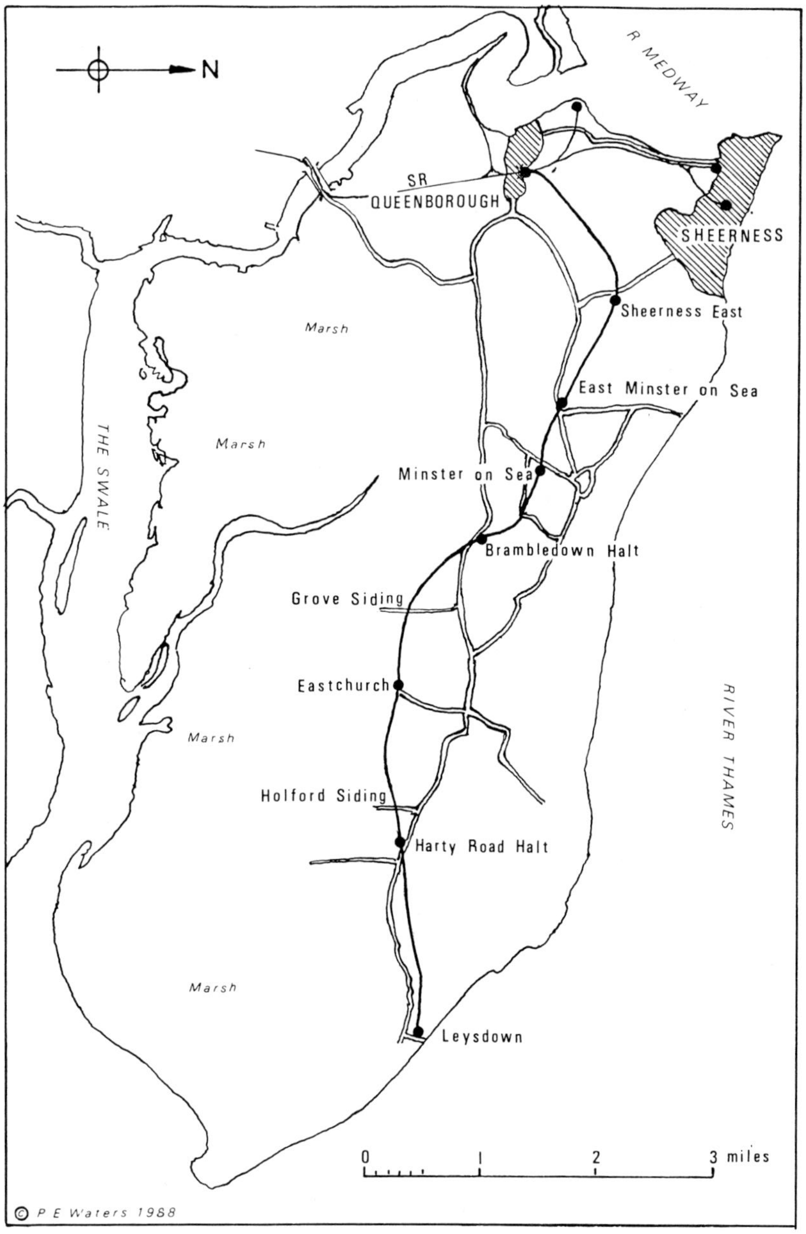

98. A rare picture of Kitson built steam railcar No. 1 in use at Queenborough c. 1901. These railcars were withdrawn and stored during World War 1 after which the bodies were mounted in pairs on a common articulated centre bogie. They were then re-used on the Sheppey, locomotive hauled.

99. East Minster with its timber lattice fencing, corrugated iron shelter and timber-faced platform c. 1901.

100. Minster on Sea with its corrugated iron station building with Stephens's style awning and basic lighting (small oil lamps) c. 1901, possibly before the opening. Note the group on the pump trolley, who look like an inspection party.

101. A basic but neat halt at Harty Road with wooden platform and shelter. This was opened in 1905 and was of a different design to the Colonel Stephens period stations.

102. Leysdown Station shortly after the opening showing the original timber platform facing and original lattice fencing. The corrugated iron station building is of a different style to those of the previous Stephens's lines and larger.

103. Journey's end: Leysdown Station now with concrete faced platform, iron railings and gas lamps.

7

The Plymouth, Devonport and South Western Junction Railway

The 9.5 mile standard gauge branch from Bere Alston to Callington was opened on 2 March 1908 and used, in part, the track bed of the 3 ft 6 in gauge East Cornwall Mineral Railway. Callington to Gunnislake was closed on 5 November 1966, but the rest of the line still provides a much needed passenger service to an area of difficult road access. Stephens was employed to regauge and reconstruct the East Cornwall Mineral Railway and also to assist with the new construction as is shown by the typical Stephens stations on the new construction and by the choice of rolling stock. He was also Locomotive Superintendent and General Manager for a short time, but his other railway commitments prevented him from continuing to hold those positions.

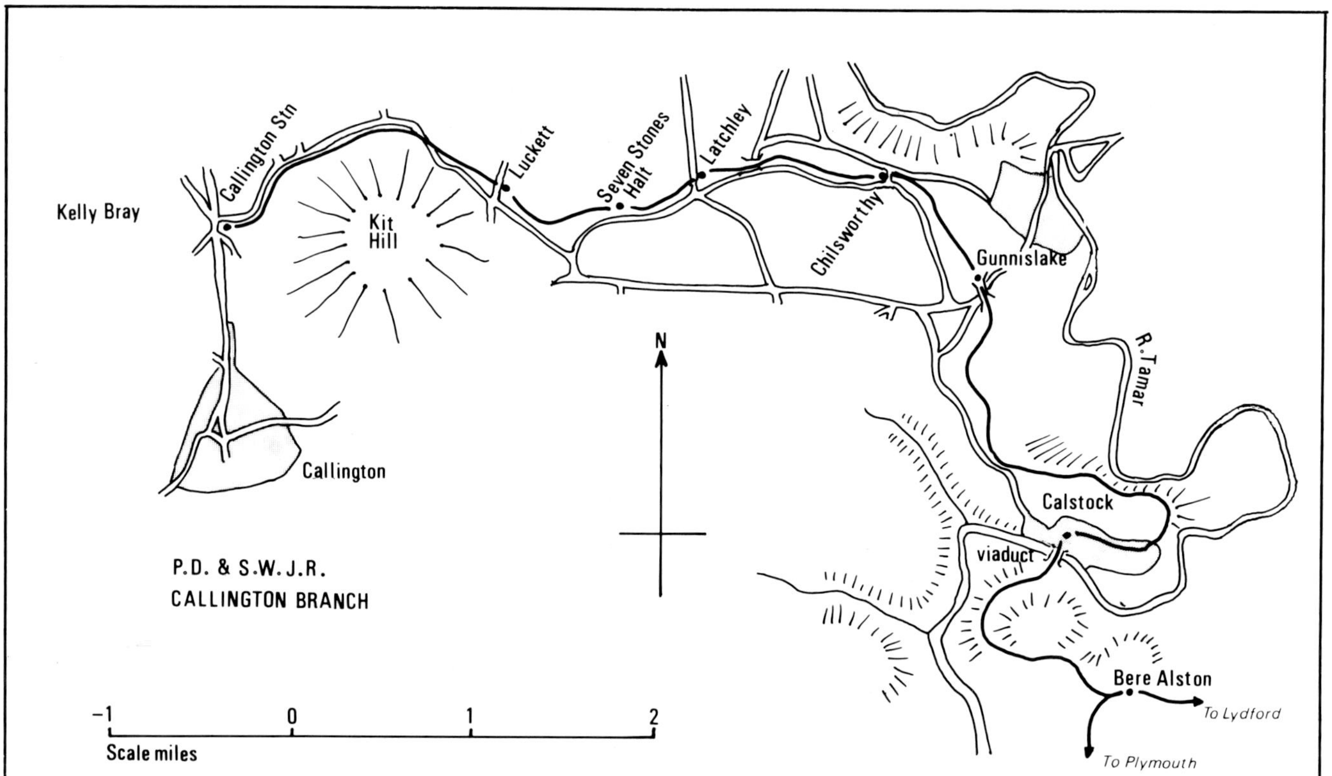

104. 0-6-0T, *A. S. Harris*, in the branch platform at Bere Alston on a mixed train containing ex-LSWR carriages c. 1910.

105. Calstock Viaduct was designed by Messrs. Galbraith and Church, the PD&SWJR's engineers. Stephens used to say that the foundations were as deep in the ground as the viaduct was tall.

106. 0-6-0T, *A. S. Harris*, and a train of ex-LSWR 4-wheeled stock crosses Calstock Viaduct c. 1910 (reputedly on the opening day). Note the wagon lift at the far end of the viaduct which served the river quay and replaced the rope-worked incline of the East Cornwall Mineral Railway.

107. Calstock Viaduct on 10 April 1977, showing the remains of the girders for the wagon lift.

108. Calstock Station on 10 April 1977, showing the typical Stephens style building. From here there is no indication of the deep river gorge which lies immediately behind the plate layers hut and van body at the end of the yard.

109. Hawthorne Leslie built 0-6-0T, *A. S. Harris*, on a mixed train climbing the 1 in 40 gradient from Calstock to Gunnislake c. 1910. At this time the PD&SWJR locomotives were painted Oxford blue like most of Stephens's other locomotives.

110. Gunnislake Station on 24 May 1908 showing the island platform and a large, double sided version of Stephens style building. Note the third rail in the foreground for the 3 ft 6 in gauge. It is recorded that no more than two days traffic were lost during the gauge conversion and this photograph illustrates how that was achieved.

111. 0-6-2T No. 5 at Chilsworthy station c. 1910.

112. Another view of Chilsworthy showing the details of the station including the two arm slotted post signal.

113. 02 0-4-4T No. 30225 at Callington on 18 August 1954 with a train consisting of a long wheelbase 4-wheel Maunsell utility van and two ex-LSWR bogie brake thirds.

114. 02 0-4-4T No. 30225 earlier the same day under the overall station roof.

8

The Burry Port and Gwendraeth Valley Railway

This 13 mile (from Burry Port to Pontyberem) standard gauge line was originally opened in stages from July 1869 for freight only. Stephens rebuilt and converted the line for passenger traffic which commenced on 2 August 1909. It was closed to passengers on 20 September 1953, but is still partly open for freight.

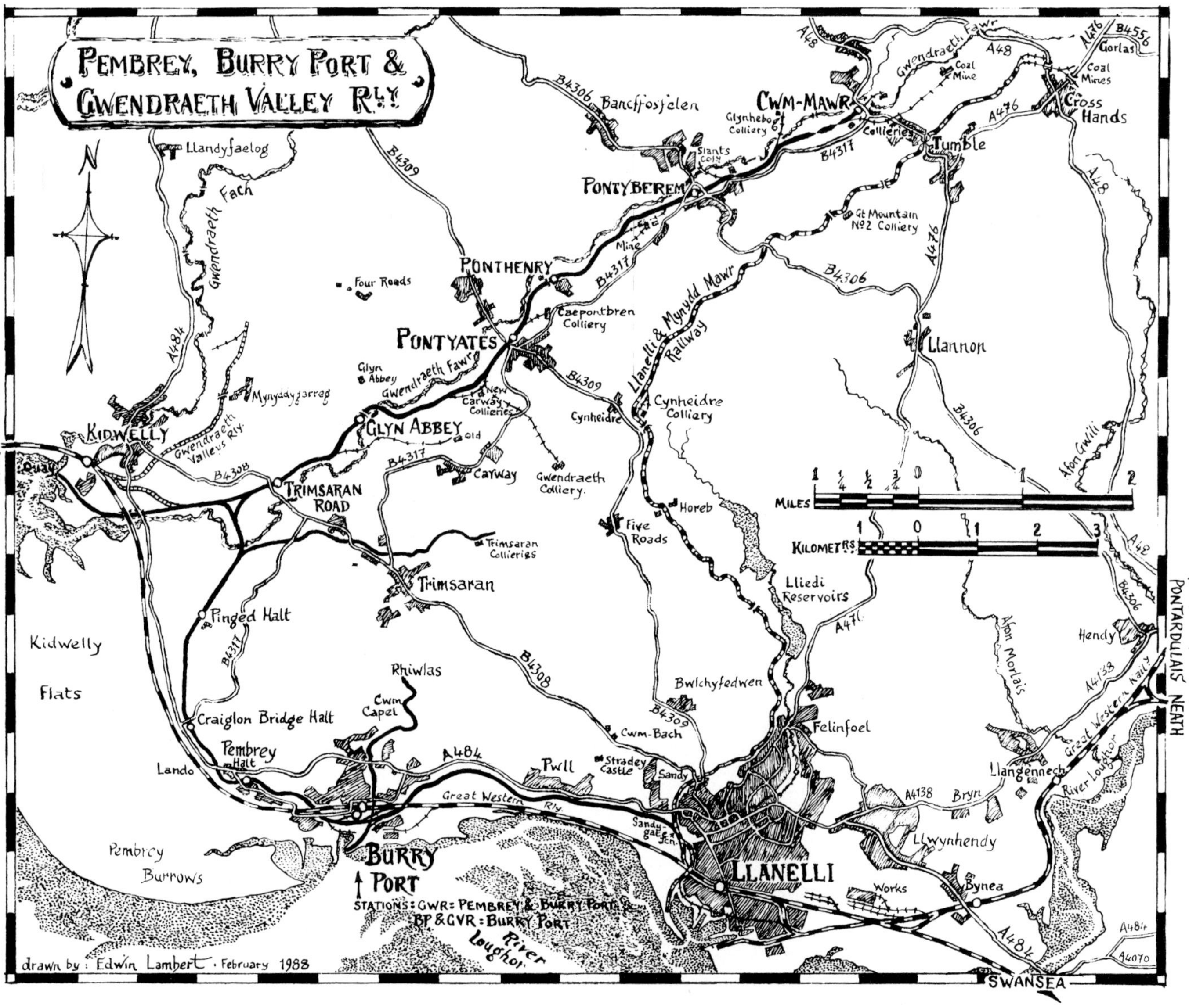

115. Burry Port on opening day, 2 Aug 1909, with Col Stephens (far right, holding coat) and members of the line's staff gathered around a Hudswell 0-6-0T.

116. The docks at Burry Port showing the coal chutes and the lock gates at the entrance to the port c. 1910.

117. One of the first public trains on 2nd Aug 1909, with crowds of jubilant locals out for a trip up the mining valley.

118. Burry Port station with an Avonside 0-6-0ST and two ex-Metropolitan Railway vehicles c. 1910. Like the PD&SWJR, a somewhat larger than previously version of the typical Stephens corrugated iron building was provided.

119. Another view of Burry Port station showing a siding full of 4-wheeled and ex-Metropolitan Railway stock c. 1910.

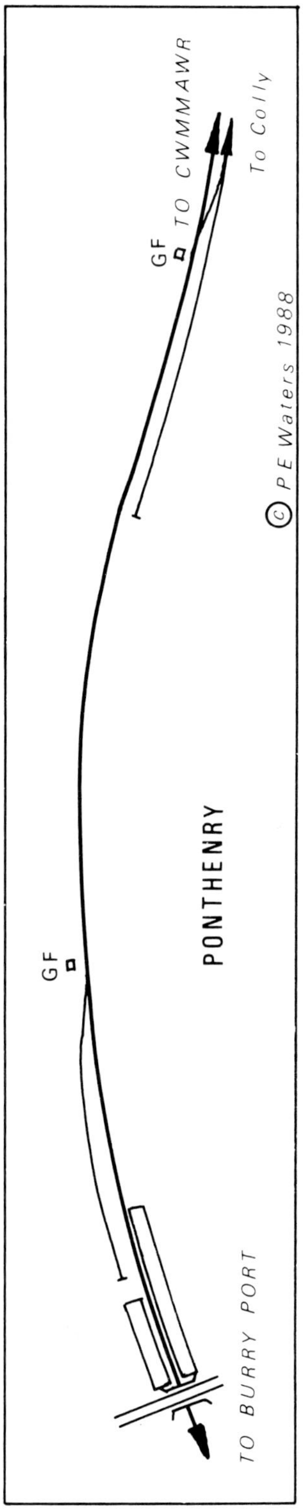
PONTHENRY
GF
GF
TO CWMMAWR
To Colly
TO BURRY PORT
© P E Waters 1988

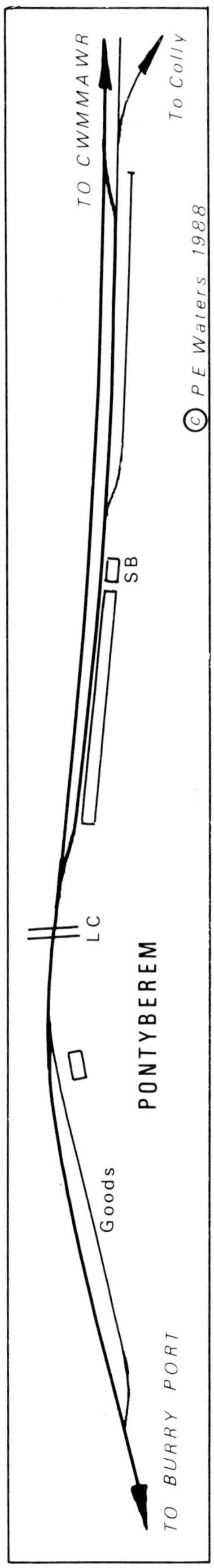
PONTYBEREM
TO CWMMAWR
To Colly
SB
LC
Goods
TO BURRY PORT
© P E Waters 1988

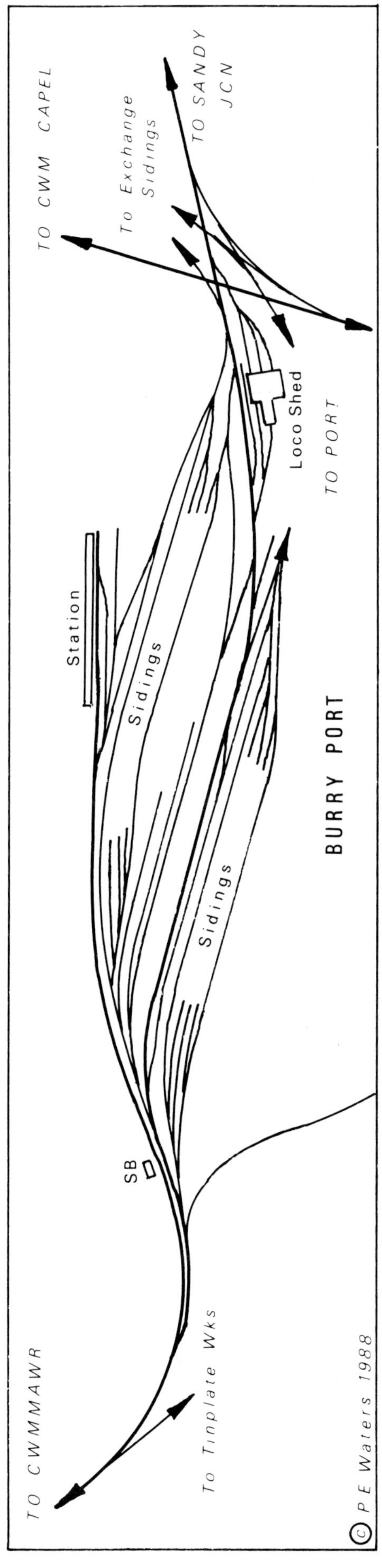
BURRY PORT
TO CWM CAPEL
To Exchange Sidings
TO SANDY JCN
Loco Shed
TO PORT
Station
Sidings
Sidings
SB
TO CWMMAWR
To Tinplate Wks
© P E Waters 1988

120. Ponthenry station with a group of locals c. 1910. Note the brick construction and the two platforms along each side of the single track.

121. Pontyberem station with an Avonside 0-6-0ST and a train of ex-Metropolitan eight-wheeled stock entering the platform in 1909.

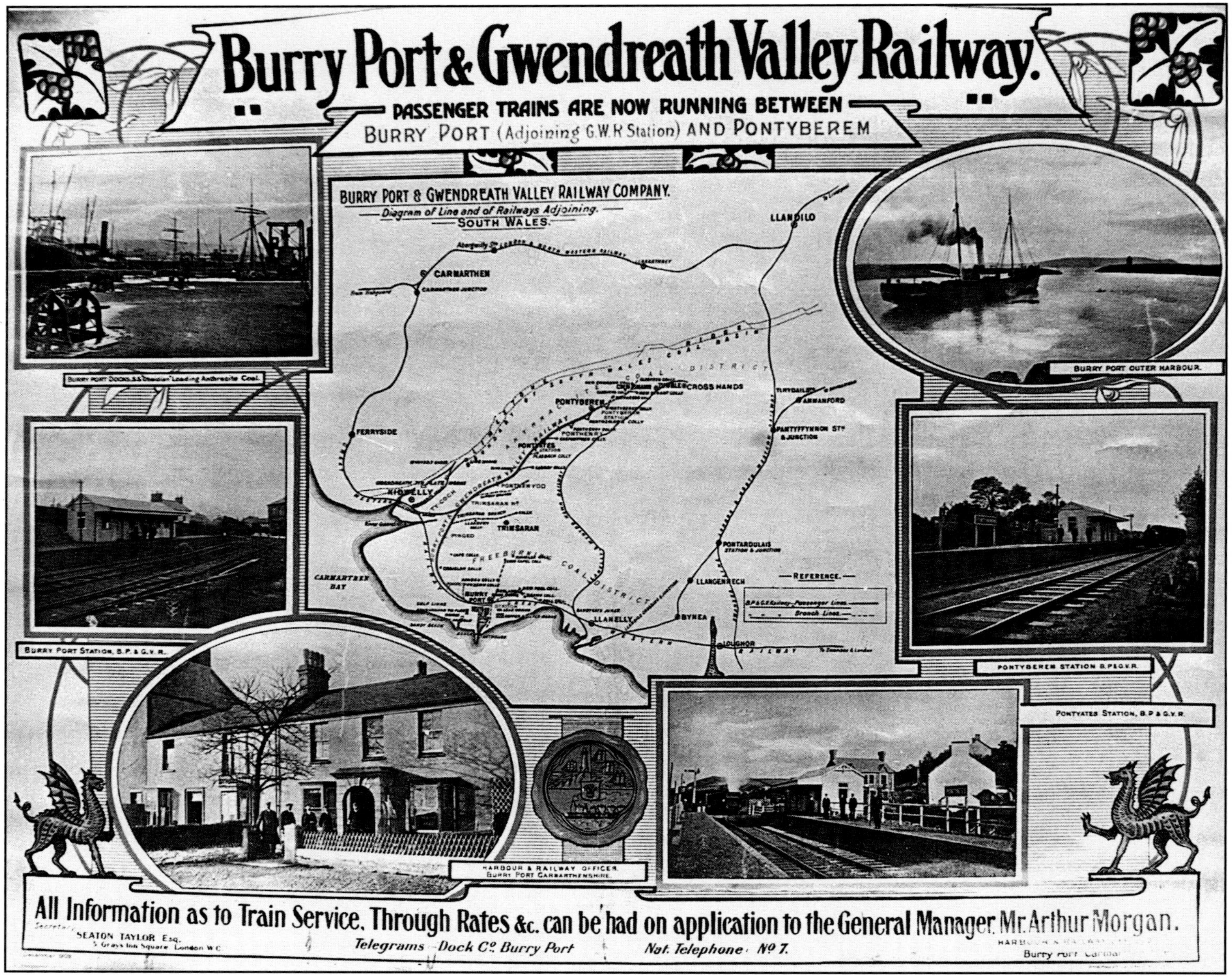

122. A Burry Port & Gwendreath Valley Railway poster dated December 1909 and announcing the new passenger service.

9

The Shropshire and Montgomeryshire Railway

The S.& M. was originally opened as the Potteries, Shrewsbury and North Wales Railway in 1866 from Shrewsbury to Blodwell Junction and closed in 1880. Stephens reconstructed it as a standard gauge light railway and it was reopened from Shrewsbury to Llanymynech, 20 miles, on 13 April 1911 and to Criggion, a further 7 miles, on 4 January 1912. It closed again to passengers by the War Department in 1941 and remained under military control until final closure of the whole railway in 1960. Stephens was Engineer, Locomotive Superintendent and Managing Director.

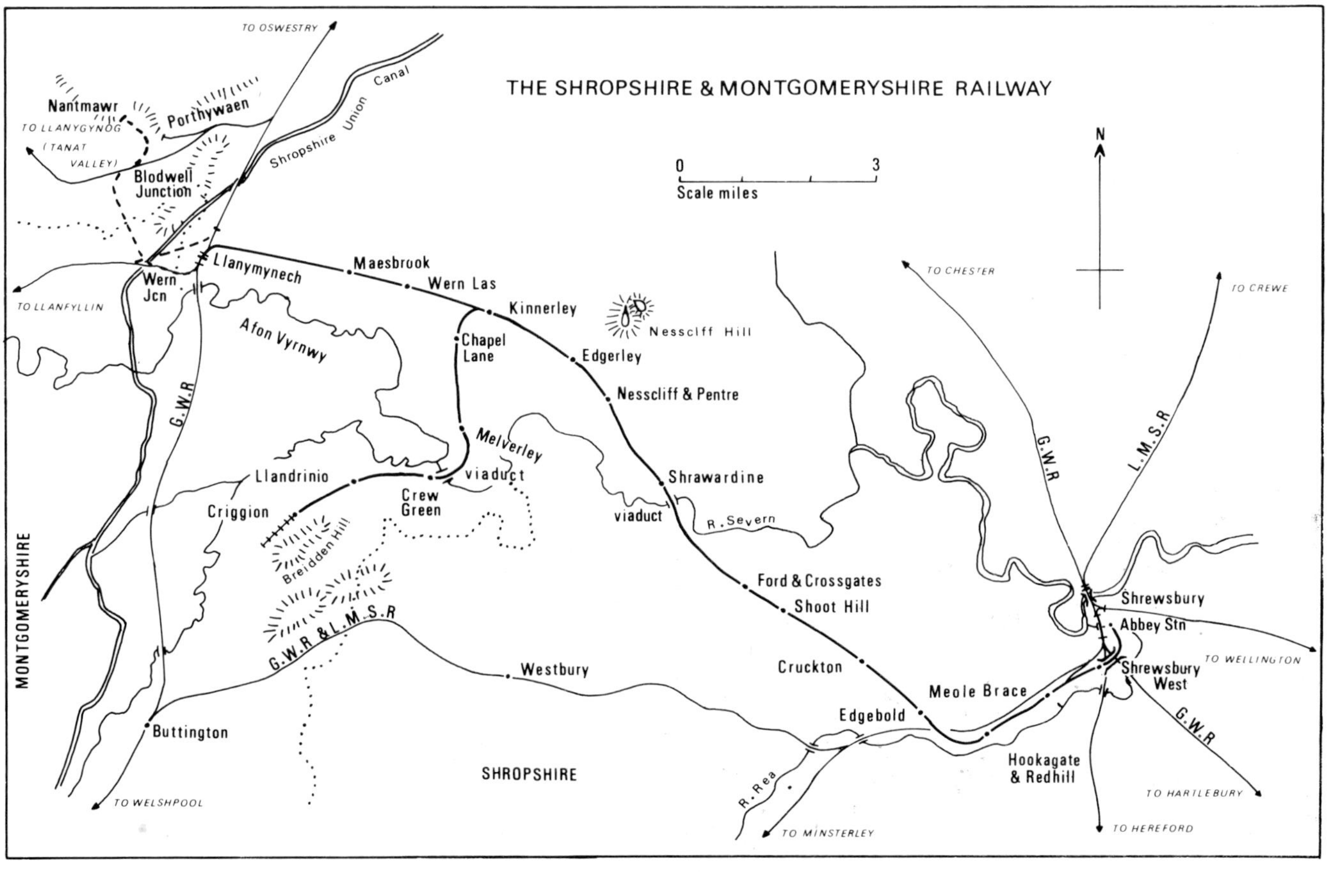

123. Ex-LNWR 0-6-0 and train of ex-Midland stock at Abbey Station in the early 1930's. Note the lorry at the far end of the station and the S&MR cattle van at the right. Although not clear from the photograph, the locomotive retains its LMS livery and cab-side number 8108.

124. Ilfracombe Goods 0-6-0 No. 3, *Hesperus*, in the platform at Abbey station with an ex-Midland 4-wheeled full brake parcels van in the mid-1930's.

125. Meole Brace Station with its timber platform and building under an over bridge.

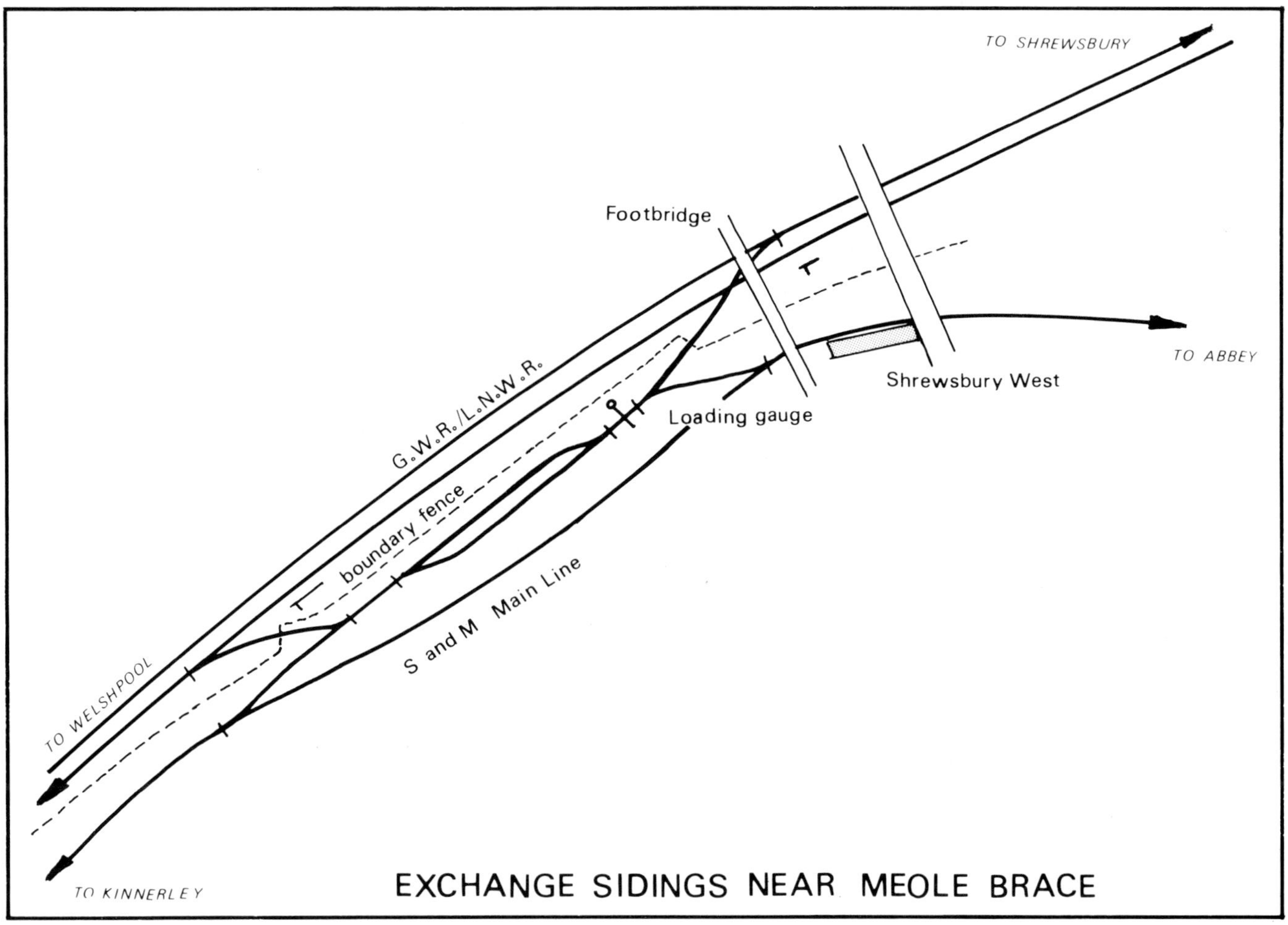

126. The timber platformed station at Hookagate with its boarded-up building. The only creature comfort seems to be a single bench on the platform c. 1935

127. Edgebold Station with original PS&NWR stone built platform, 1911 wooden building and two-bar timber fencing c. 1930.

128. Cruckton Station, with brick built platform covered in weeds and wild flowers c. 1930. A truly lonely spot to wait for any infrequent trains.

129. Shoot Hill Station with another brick built platform and an old van body as its only building c. 1950.

130. Ford & Crossgates Station with its original PS&NWR stone built platform and refurbished wooden building, ex-Midland van body and wooden bench on the platform.

131. The great iron and stone PS&NWR viaduct at Shrawardine with its details clearly shown c. 1930. The girders from the disused second track were probably reused in the Melverley Viaduct, which was rebuilt by Stephens in 1911.

132. The sleeper built platform at Edgerley.

133. The view east from the road bridge at Kinnerley showing an ex-LNWR 0-6-0 and ex-Midland 4-wheeled van c. 1935.

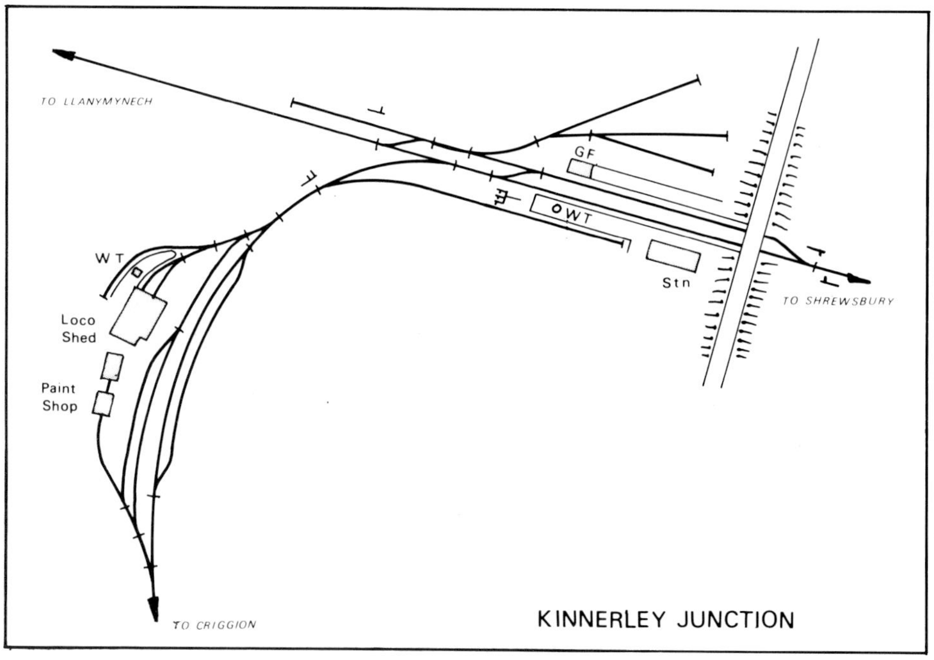

134. Kinnerley station looking east with a train in the platform headed by an Ilfracombe Goods 0-6-0 on a busy day c. 1920. A magnifying glass reveals another train disappearing into the distance towards Shrewsbury. Note the detail shown in this picture: the ground frame, platform lamps, etc.

135. A general view of Kinnerley station from below the overbridge, showing the original PS&NWR brick building with the raised platform for milk churns. Note the water tower and three arm bracket signal for the junction at the far end of the platform c. 1935.

136. Ex-LNWR 0-6-0 No. 8182 still in LMS livery taking water at Kinnerley. Note the raised lettering on the ex-Midland bogie coach.

137. A general view of Kinnerley shed and works with two of the ex-LNWR 0-6-0's on shed and a Terrier in the storage siding c. 1935. To the right can be seen an Ilfracombe Goods cab, ex-LCC horse tram car and the Wolseley-Siddeley railcar.

138. A close-up of the locomotive shed c. 1936, with ex-LNWR 0-6-0 No. 8236 still carrying its cast LMS number plate on the smokebox door.

139. 0-4-2WT, *Gazelle*, together with its trailer consisting of the body of the Wolseley-Siddeley railcar mounted on the underframe of the ex-LCC horse tram, seen c. 1937 after their rebuild as the line's inspection train.

140. Stored in the far siding next to the Shed stand the Ford railbus set, *Gazelle* and some wagons c. 1940.

141. Derelict Ford railbus set at Kinnerley in the early 1930's. Note the tender of the second Ilfracombe Goods behind and the engine and gearbox dumped at the right.

142. Ex-LNWR 0-6-0 No. 8236, with its front supported on blocks and its cylinder covers off, at Kinnerley during the period of military control. Note the Hunslet diesel in the background.

143. Maesbrook with its original PS&NWR platform and crossing keepers house, 1911 wooden station building and War Department continental style level crossing barriers c. 1950.

144. Another view of Maesbrook station showing the hand operated diamond request stop signal in the late 1930's.

145. Llanymynech showing the original PS&NWR platforms and refurbished station building with ex-LNWR 0-6-0 and train of ex-Midland stock arriving from Shrewsbury on 28 May 1932.

146. Ford railbus set at Llanymynech on 27 July 1931, looking towards Kinnerley.

147. Not a potential passenger at Chapel Lane station in the 1930's. This stone built platform was one of the five stations on the Criggion branch which ran from Kinnerley to Criggion Quarry. This branch was not taken over by the military in 1941 and continued to be operated by the original company. The quarry used a 4-wheeled Sentinel shunter on its trains.

148. Melverley station in the 1930's. The distinctive road bridge, platform and building are all of PS&NWR origin.

149. Criggion Station at the end of the branch. Note *Gazelle* and coach at the end of the platform forming, apparently, the Birmingham Locomotive Club special train on 23 April 1939.

10

The Weston, Clevedon and Portishead Light Railway

This line was originally opened from Weston-super-Mare to Clevedon on 1 December 1897 as the Weston Clevedon and Portishead Tramway. It was opened to Portishead as a light railway on 7 August 1907, making a total of 14.25 miles of standard gauge. It closed entirely in October 1940. Stephens became involved in 1911 as traffic manager, but rapidly gained control of the Company to become Engineer, Locomotive Superintendent and General Manager.

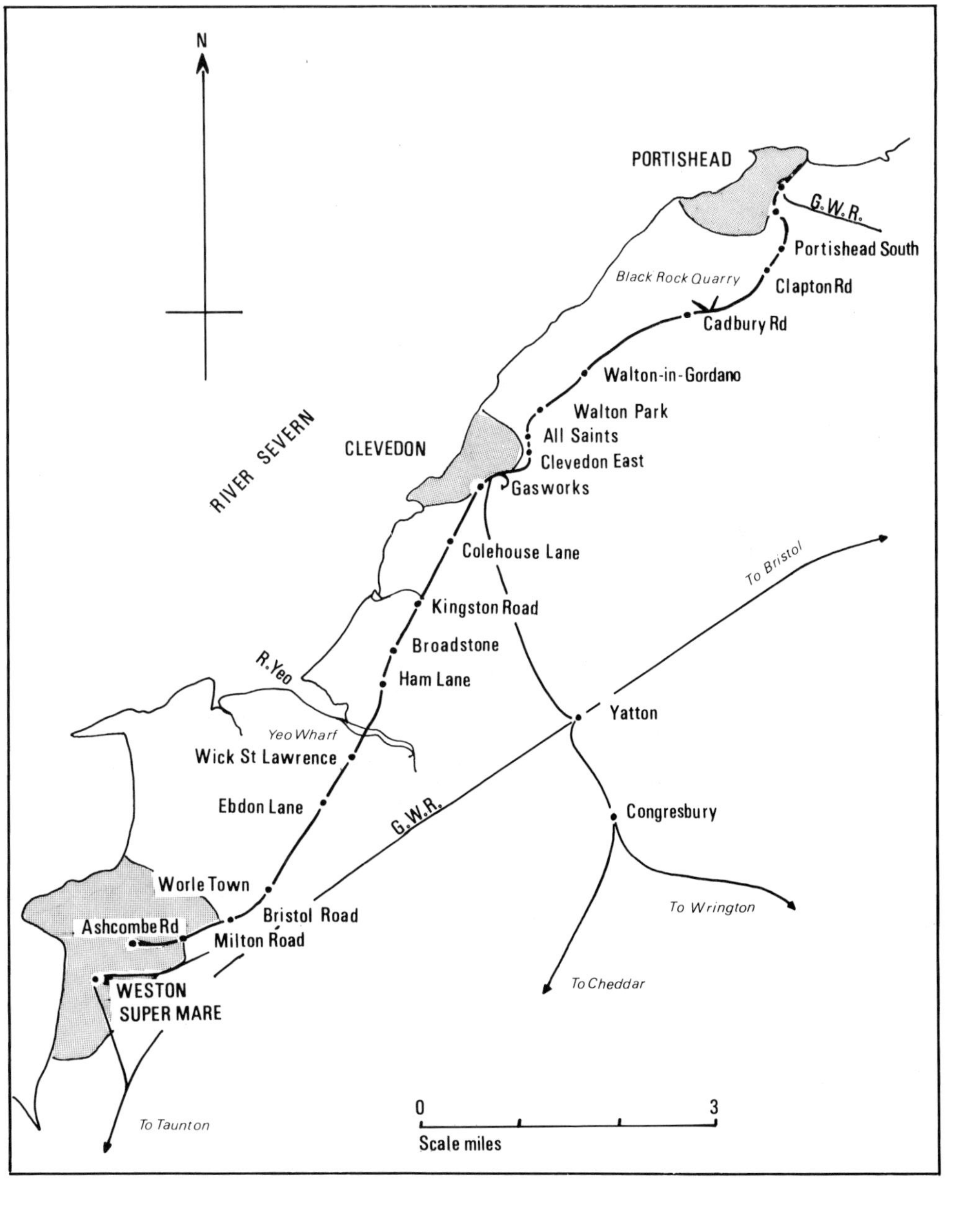

150. A Victorian period scene c. 1899 at Weston with one of the American styled balcony-end carriages in the station, next to which is the Lewisham House School "Playing Ground" and a substantial stone town house.

151. A later view of Weston Station (the Playing Ground is now the Lewisham House Sports Field) showing the light track and ground level platform.

152. Ex-GWR Sharp Stewart 2-4-0T, later named *Hesperus* at Weston with one of the American style bogie coaches c. 1920.

153. Manning Wardle 0-6-0ST, *Weston*, at Weston c. 1930.

154. Holiday crowds at Weston in the early 1930's.

155. Terrier No. 2, *Portishead*, waits at Weston c 1937 with a train of ex-LSWR and ex-Metropolitan 4-wheeled stock.

156. Basic facilities at Worle Town c. 1930 with the small railcar just visible on the right.

157. Cole House Lane station in the 1930's.

158. 0-6-0ST, *Harold*, on hire from the line's contractor to the Tramway Company at Clevedon c. 1897. Note the light track and half round sleepers in the foreground.

159. The station building at Clevedon on 25 June 1938.

160. Terrier No. 4 again at Clevedon on a Portishead train consisting of two ex-Metropolitan 4-wheeled carriages on 25 June 1938.

161. No. 4, the second Terrier, in full Southern style green lined livery at Clevedon on 4th October 1937 shortly after purchase. The 4-wheeled carriage is ex-Taff Vale and the second vehicle is the ex-Southern Railway Drewry railcar.

162. Manning, Wardle 0-6-0ST No. 5 at Clevedon c. 1936. In the shed in the background can be seen the ex-Southern Railway Drewry railcar.

163. Clevedon shed c. 1935, with ex-Jersey Railway 2-4-0T, *Clevedon*, (formerly *General Don*) in store at the left, ex-LSWR carriage set centre, LMS van and private owner wagon right. The vehicle in the foreground is the baggage trolley for use with the railcars.

164. Another general view of Clevedon shed on 25 June 1938 showing Terrier No 4. at the coaling stage, small Drewry railcar and ex-LSWR carriages.

165. Manning Wardle 0-6-0ST, *Weston*, in store at Clevedon shed c. 1930.

166. Ex-GWR 2-4-0T No. 4, *Hesperus*, in the shed at Clevedon c. 1930.

167. Terrier *Portishead* in the street at Clevedon with a train of ex-Metropolitan and ex-LSWR stock c. 1937.

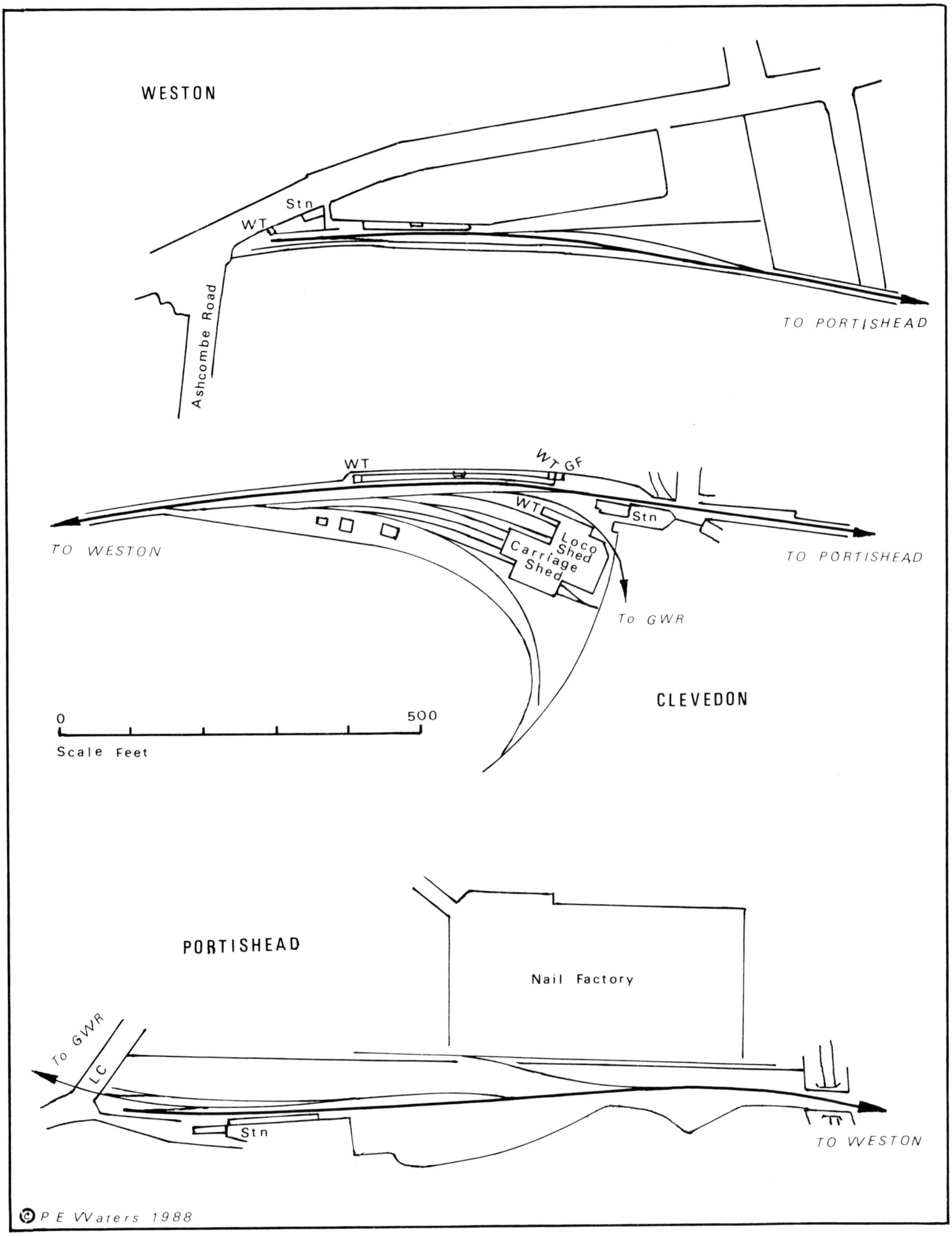
WESTON
Stn
WT
Ashcombe Road
TO PORTISHEAD
WT
WT GF
WT
Stn
Loco Shed
Carriage Shed
TO WESTON
TO PORTISHEAD
To GWR
CLEVEDON
0
500
Scale Feet
PORTISHEAD
Nail Factory
To GWR
LC
Stn
TO WESTON
© P E Waters 1988

168. All Saints Clevedon in the 1930's, consisting of just a name and timetable board.

169. Walton Park c. 1930, with small shelter and gas lamp.

170. Manning Wardle 0-6-0ST *Weston* with a train of the American style bogie vehicles at Walton-in-Gordano Station c. 1910. The shelter, although looking like a Stephens building, was actually constructed before he obtained control of the line.

171. Portishead South in the 1930's with a similar building to that at Walton-in-Gordano.

172. An early view of an unidentified level crossing (note the WC&PT notice) showing the rather basic construction.

173. Manning Wardle 0-6-0ST No. 5 shunts four 6-plank private owner wagons at Portishead c. 1925. Note the disk wheels, said to have been ordered by Stephens in the belief that they would improve adhesion.

174. An hour later No. 5 starts its train of two ex-Metropolitan 4-wheelers and two open wagons for Clevedon.

175. Terrier, *Portishead,* with the ex-Metropolitan 4-wheeled set leaving Portishead for Clevedon c. 1935. The 4-wheeled stock had steps added for use at ground level platforms.

176. A line of empty ex-LSWR 4-wheeled stock stands at Portishead Station with its doors open after the holiday rush had departed c. 1929.

177. A Weston Clevedon & Portishead Railway handbill dated September 1939.

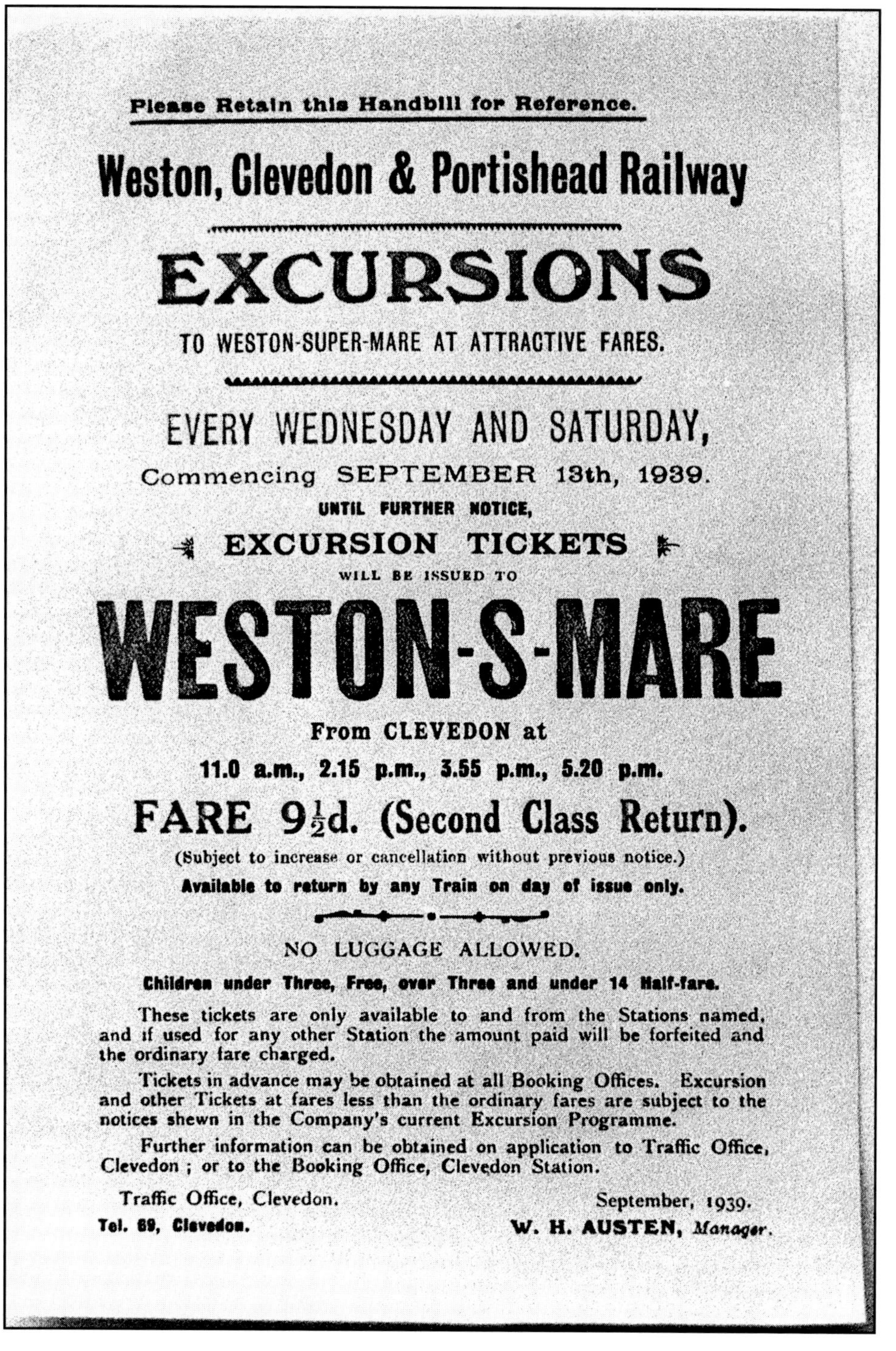

Please Retain this Handbill for Reference.

Weston, Clevedon & Portishead Railway

EXCURSIONS

TO WESTON-SUPER-MARE AT ATTRACTIVE FARES.

EVERY WEDNESDAY AND SATURDAY,

Commencing SEPTEMBER 13th, 1939.

UNTIL FURTHER NOTICE,

EXCURSION TICKETS

WILL BE ISSUED TO

WESTON-S-MARE

From CLEVEDON at

11.0 a.m., 2.15 p.m., 3.55 p.m., 5.20 p.m.

FARE 9½d. (Second Class Return).

(Subject to increase or cancellation without previous notice.)

Available to return by any Train on day of issue only.

NO LUGGAGE ALLOWED.

Children under Three, Free, over Three and under 14 Half-fare.

These tickets are only available to and from the Stations named, and if used for any other Station the amount paid will be forfeited and the ordinary fare charged.

Tickets in advance may be obtained at all Booking Offices. Excursion and other Tickets at fares less than the ordinary fares are subject to the notices shewn in the Company's current Excursion Programme.

Further information can be obtained on application to Traffic Office, Clevedon; or to the Booking Office, Clevedon Station.

Traffic Office, Clevedon. September, 1939.

Tel. 69, Clevedon. W. H. AUSTEN, *Manager*.

11

The East Kent Railways

This standard gauge line was opened in sections for coal and goods traffic from November 1912 and for passenger traffic from Shepherdswell to Wingham on 16 December 1916, and to Canterbury Road, 11.25 miles from Shepherdswell, and to Sandwich Road, 2.5 miles from the junction at Eastry, in 1925. The passenger service between Eastry and Canterbury Road was withdrawn on 31 October 1928 and from all other stations on 30 October 1948. The goods service between Eastry and Richborough was withdrawn on 27 October 1949 and between Eastry and Canterbury Road on 25 July 1950. Eythorne to Eastry was closed on 1 July 1951, leaving the line from Shepherdswell to Tilmanstone Colliery to survive to the 1980's. Stephens was the Engineer of all of these lines.

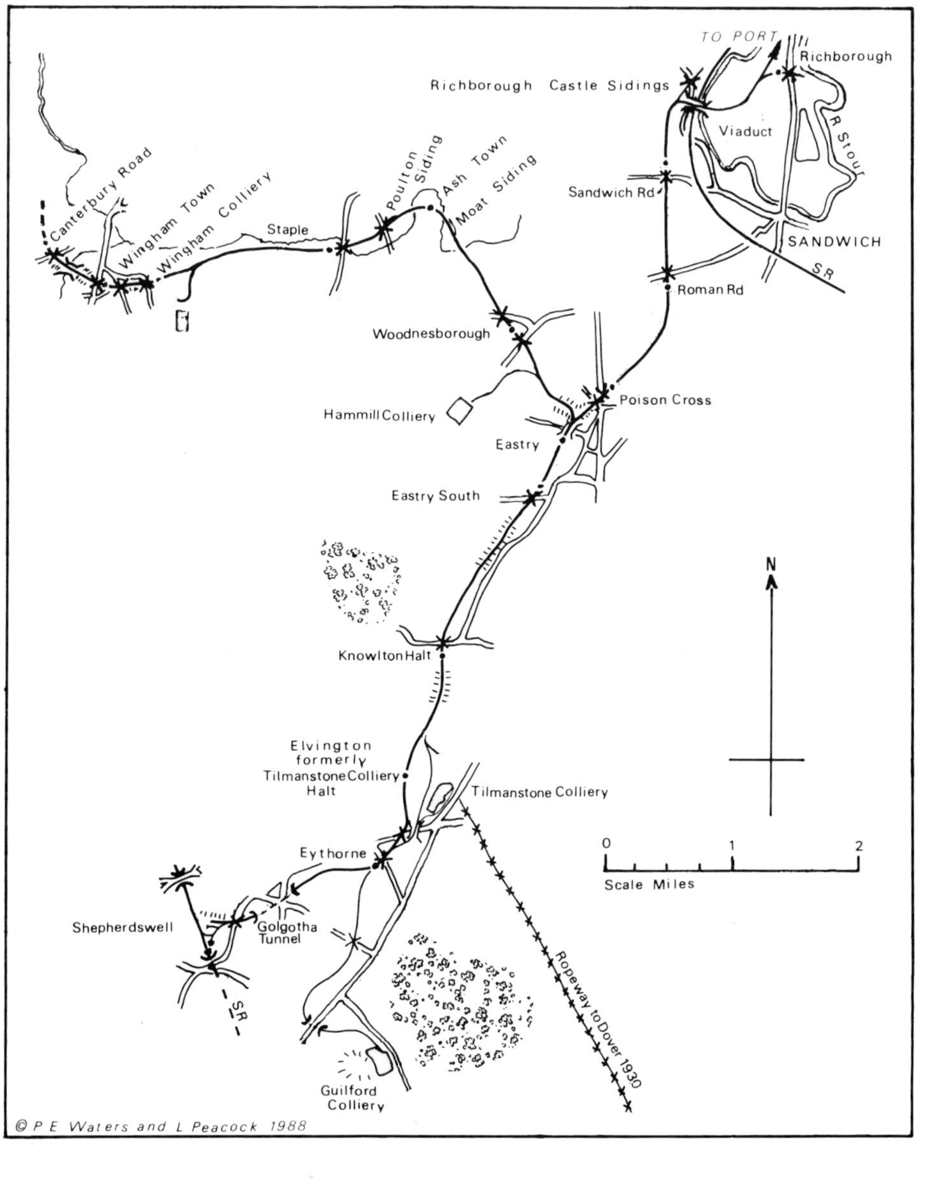

178. A crowded scene at Shepherdswell Station in the late 1920's, showing ex-LSWR and ex-LC&DR carriages, ex-LC&DR van and various open wagons in the platform road. An ex-LC&DR brake-3rd and six wheelers are in the adjacent siding and the locomotive shed can be seen in the background. Note the station buildings including the round hut.

179. Shepherdswell Station looking towards the stop blocks c. 1930. The hut on the hill was used as living accomodation by some of the staff.

180. Another typically crowded scene at Shepherdswell in the 1940's.

181. A general view of Shepherdswell yard showing the shed and dump sidings c. 1946.

182. Shepherdswell shed showing two of the ex-SER 0-6-0's, O/O1 No. 6 and O1 No. 100, No. 4 is in the shed and the cab and frames of No. 1 can be seen at the left c. 1938.

183. Shepherdswell shed c. 1925 with the ex-LSWR 4-4-2T No. 5 in the foreground.

184. K&ESR 2-4-0T No. 2, *Northiam*, under overhaul at Shepherdswell in the 1920's. This locomotive travelled afar on several occasions. She was loaned to the WC&PR and the EKR, but perhaps her most famous journey was to Basingstoke where she starred in the film "Oh! Mr. Porter".

185. K&ESR 0-8-0T No. 4, *Hecate*, on the EKR when on loan c. 1918. Note the horseshoe on the top lamp iron.

186. Ex-SER 0-6-0 No. 6 and hand crane at Shepherdswell on 30 June 1934.

187. K&ESR No. 2, *Northiam,* at Shepherdswell with ex-GWR 0-6-0ST EKR No. 1, on 14 May 1927.

188. Dumped 4-wheeled passenger stock at Shepherdswell on 18 July 1936.

189. 0-6-0 No. 6 passing through Shepherdswell yard on the way to the exchange sidings with a coal train made up of one Great Eastern five-plank wagon and a long line of SE&CR opens. Note the EKR open wagons on the bank behind.

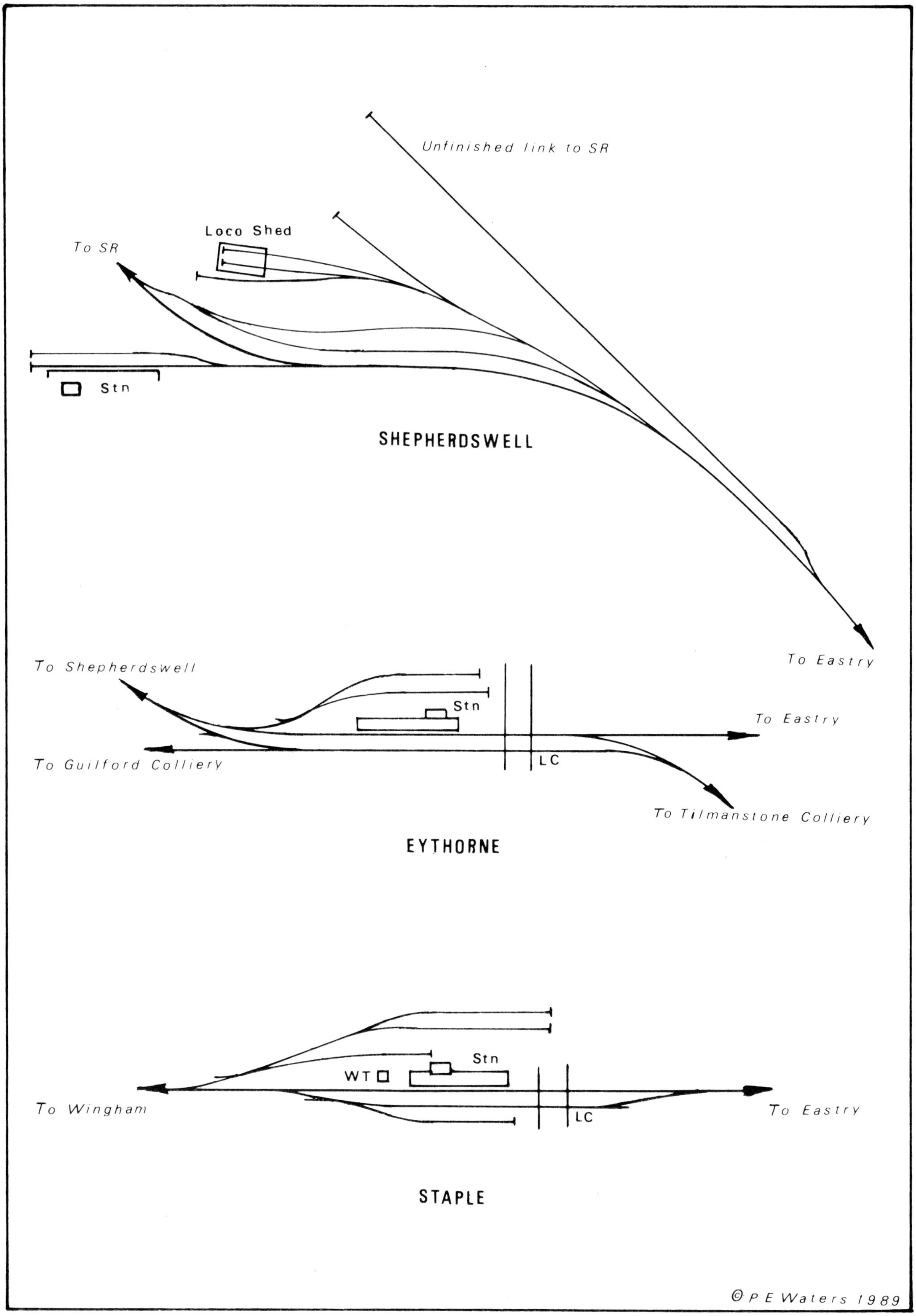

Unfinished link to SR
Loco Shed
To SR
Stn
SHEPHERDSWELL
To Eastry
To Shepherdswell
Stn
To Eastry
To Guilford Colliery
LC
To Tilmanstone Colliery
EYTHORNE
Stn
WT
To Wingham
LC
To Eastry
STAPLE
© P E Waters 1989

190. South portal of Golgotha Tunnel on 3 October 1987 showing the double track construction with only a single track laid. A long block of unexcavated chalk about 8 feet high occupied the second road.

191. Guilford Colliery c. 1917. This was not a successful pit and had closed by 1921.

192. Ex-WC&PR 0-6-0ST No. 2, *Walton Park*, with the ex-K&ESR bogie brake third at Eythorne c. 1925.

193. Eythorne Station after the passenger trains had gone c. 1955. Note the brick built station and crossing keepers hut.

194. Tilmanstone Pit Head c. 1920.

195. An interesting aerial view of Tilmanstone Colliery. Note the standard gauge sidings at the back of the mine and the two sets of pit-head gear visible beyond the cloud. Locomotive Magazine said, "The extensive colliery sidings at Tilmanstone, laid out by Mr Stephens, are provided with up-to-date electrically-driven screens and six roads and well worth a visit. Shunting has been facilitated by the construction of a hump. . ." Note also the narrow gauge lines in the mine area and the footbridge across the sidings leading to a path to Elvington Halt and Village.

196. Eastry South looking towards Eastry c. 1930.

197. Eastry Station about 1930 showing the junction of the Canterbury Road line to the left and the Richborough line to the right. Note the railings of the under bridge beyond the turnout.

198. No 2 with an ex-LC&DR coach on the bridge at Eastry c. 1938.

199. Woodnesborough Station showing the small timber building and brick platform c. 1936.

200. In the afternoon sunshine in 1946, 0-6-0 No. 6 and an ex-LSWR bogie brake 3rd stand at Staple with a train for Shepherdswell.

201. Staple looking towards Wingham c. 1930. Note the wind pump by the water tank.

202. Wingham Town station c. 1925 with locomotive No. 2 and an ex-LSWR 6-wheeled brake third.

203. 0-6-0 locomotive No. 3 at Wingham Town in the late 1920's.

204. 0-6-0 No. 6 at Wingham with an ex-LSWR bogie corridor brake third heading towards Shepherdswell on 25 April 1947.

205. 0-6-0 No. 6 waits in the siding by the haystack at Canterbury Road while its coach is run past by gravity on 25 April 1947.

206. Canterbury Road Station in the mid-1930's bearing the simple name board "Wingham". The cutting continued towards Canterbury.

207. Canterbury Road station on 25 April 1947 looking towards Shepherdswell. The name board now reads "Wingham Canterbury Road".

208. Ex-LSWR 0-6-0ST No. 7 at Canterbury Road c. 1925.

209. 0-6-0 No. 8 and the ex-K&ESR coach at Canterbury road c. 1930.

210. Poison Cross Halt, the first station on the Richborough line. Note the padlocked mail boxes on the platform, one for each direction of travel! Hence one is said to have been labelled Poison-Sandwich.

211. 0-6-0ST No. 7 with ex-LSWR and ex-LC&DR 6-wheelers at Sandwich Road on 22 September 1928.

212. The bridge carrying the EKR over the SE&CR coast line. This bridge was not passed for passenger trains, which terminated at Sandwich Road.

213. Richborough Port Station which was never opened to passenger traffic because of the inadequate bridges over the SE&CR and the River Stour. The Richborough Port Railway crossed the EKR on the level between the platform and the road crossing. The line beyond the fenced-off level crossing led to a quay on the Stour.

214. Unlucky 13? The coach crushed between locomotive No. 3 on a freight and locomotive No. 1 was probably the ex-GER fourwheeler, ex-KESR No. 13, which had disappeared from the EKR by 1927. It is believed that this accident occurred in 1919, but no other details were known at the time that this was written.

12

The Edge Hill Light Railway

This 3 mile standard gauge mineral railway was opened in 1920 between the Burton Dassett exchange sidings and the iron stone workings on Edge Hill. It closed in 1925. Stephens was the engineer.

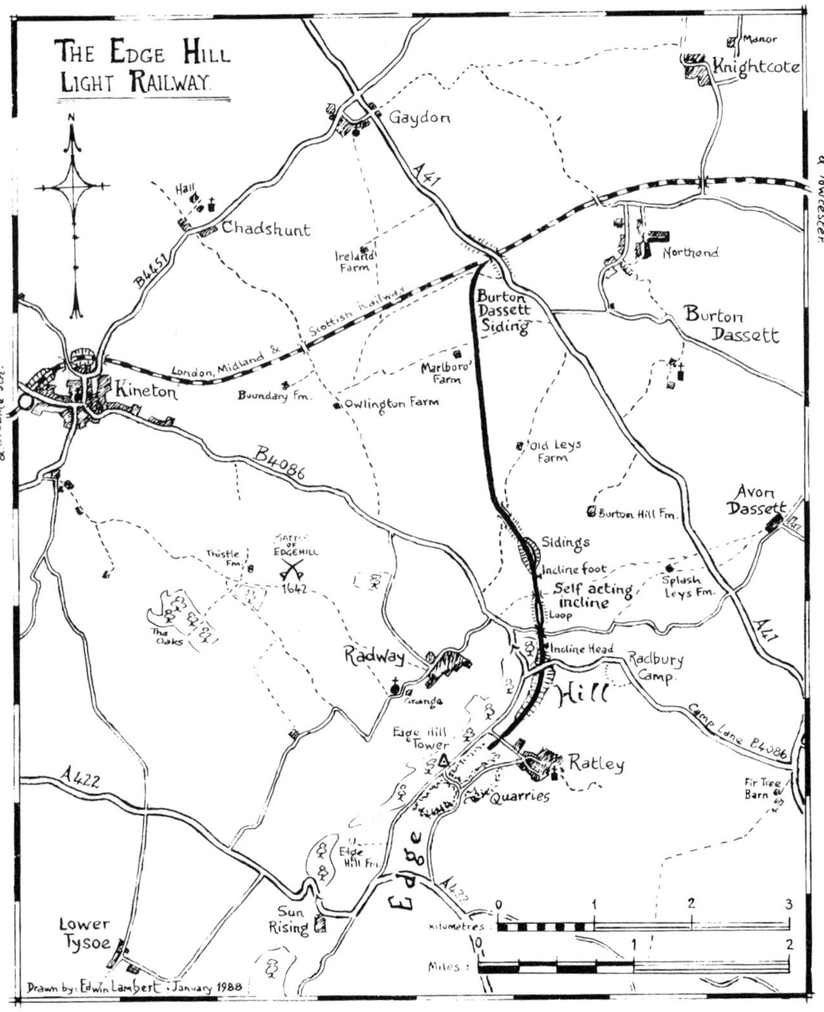

215. Perhaps the line was still working when this picture was taken. Ex-LBSCR A1X Terrier 0-6-0T No. 1 with an ex-Great Eastern brake van stands next to the makeshift water hose.

216. Both locomotives, one of the two brake vans and some derelict contractors type wagons in the sidings at the foot of the incline c. 1925.

217. 0-6-0T No. 2 and brake van No. 2 have not moved since the previous photograph, although the tarpaulin has gone from over the locomotive c. 1930. Note the incline in the background.

218. The LMS wagon sheet conceals Manning Wardle 0-4-0ST, *Sankey*, parked in front of a crude water tank wagon under the road bridge near the top of the incline c. 1930.

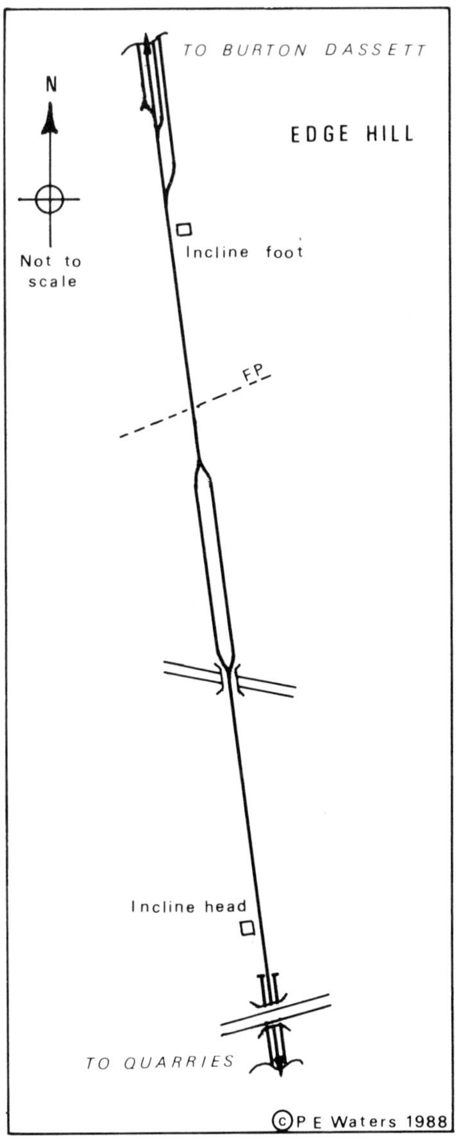

13

The Weston Point Light Railway

In 1918 Stephens had a 0.95 mile siding from the LNWR to the Castner Kellner Alkali Company's works at Weston Point, Cheshire pending approval under the Light Railway Acts. The Light Railway Order for the Weston Point Light Railway was issued on 2 Dec 1920 and the line opened on 22 May 1922. The Weston Point Light Railway is still in use for its original purpose, serving ICI's Runcorn factories.

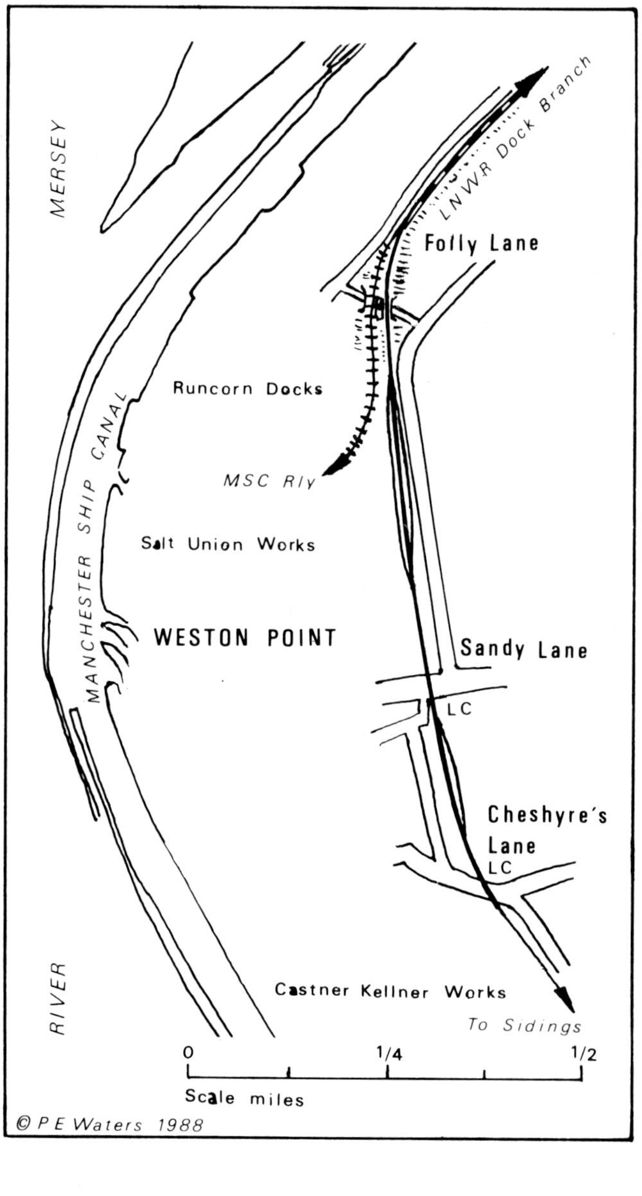

219. The end of the old LNWR Dock Branch at Folly Lane with the Weston Point Light Railway bearing away to the left. The Manchester Ship Canal Railway used to continue on ahead beyond the short siding now remaining. 4 Dec 1988.

220. Swinton's Arches showing the WPLR bridge in the foreground and the MSC Rly bridge hidden behind, with the River Mersey in the background. 4 Dec 1988.

221. View north from Sandy Lane level crossing on 4 Dec 1988.

222. Sandy Lane crossing looking south on 4 Dec 1988. As built the line had ungated crossings and flat bottom rail.

223. Cheshyre's Lane crossing looking north on 4 Dec 1988. Despite its location in the midst of ICI's Runcorn chemical plants, the WPLR still manages to retain a rural aspect.

224. ICI period enamel trespass sign on the south side of Cheshyre's Lane crossing on 4 Dec 1988. Although the WPLR was bought by ICI in 1926, it was not formally transfered to them until 1948.

14

The Snailbeach District Railways

The Snailbeach District Railways had the rather curious gauge of 2 ft 3.75 in (originally 2 ft 4 in, but apparently reduced by 1/4 in in Stephens's time) and was opened in 1873 for goods traffic only. The mineral traffic ceased during the first world war and goods traffic was virtually non-existent by 1920. A new management headed by Col Stephens took over on 1 Jan 1923. The last remnants of the line ceased to be used about 1959, but the Company still exists (the last of the companies once controlled by Stephens) and still owns the track-bed, part of which is let.

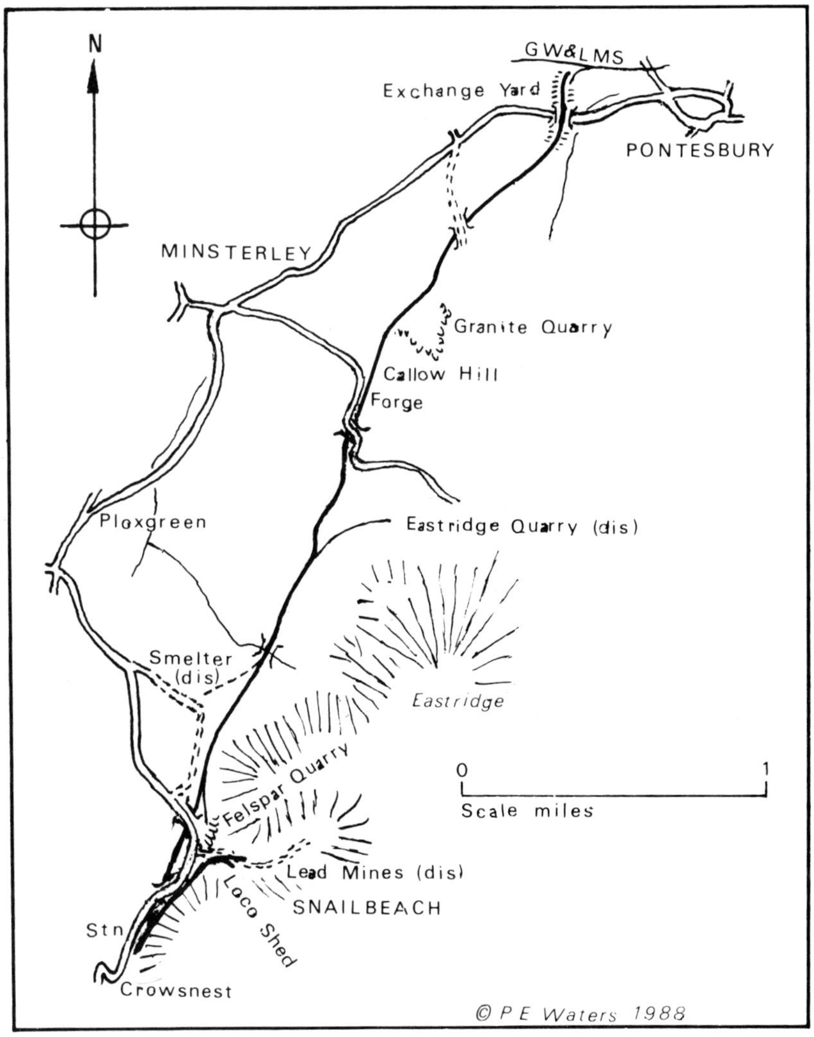

225. The exchange sidings at Pontesbury showing the transfer wharves c. 1926. Note the standard gauge line running under the wharf for gravity transfer from the narrow gauge 4-wheeled hopper wagons. Transfer in the reverse direction was facilitated by the sunken narrow gauge line and the tipping table for standard gauge wagons just discernable above and to the right of the wagon on the lower line.

226. Another view of the transfer wharves taken on the same day c. 1926.

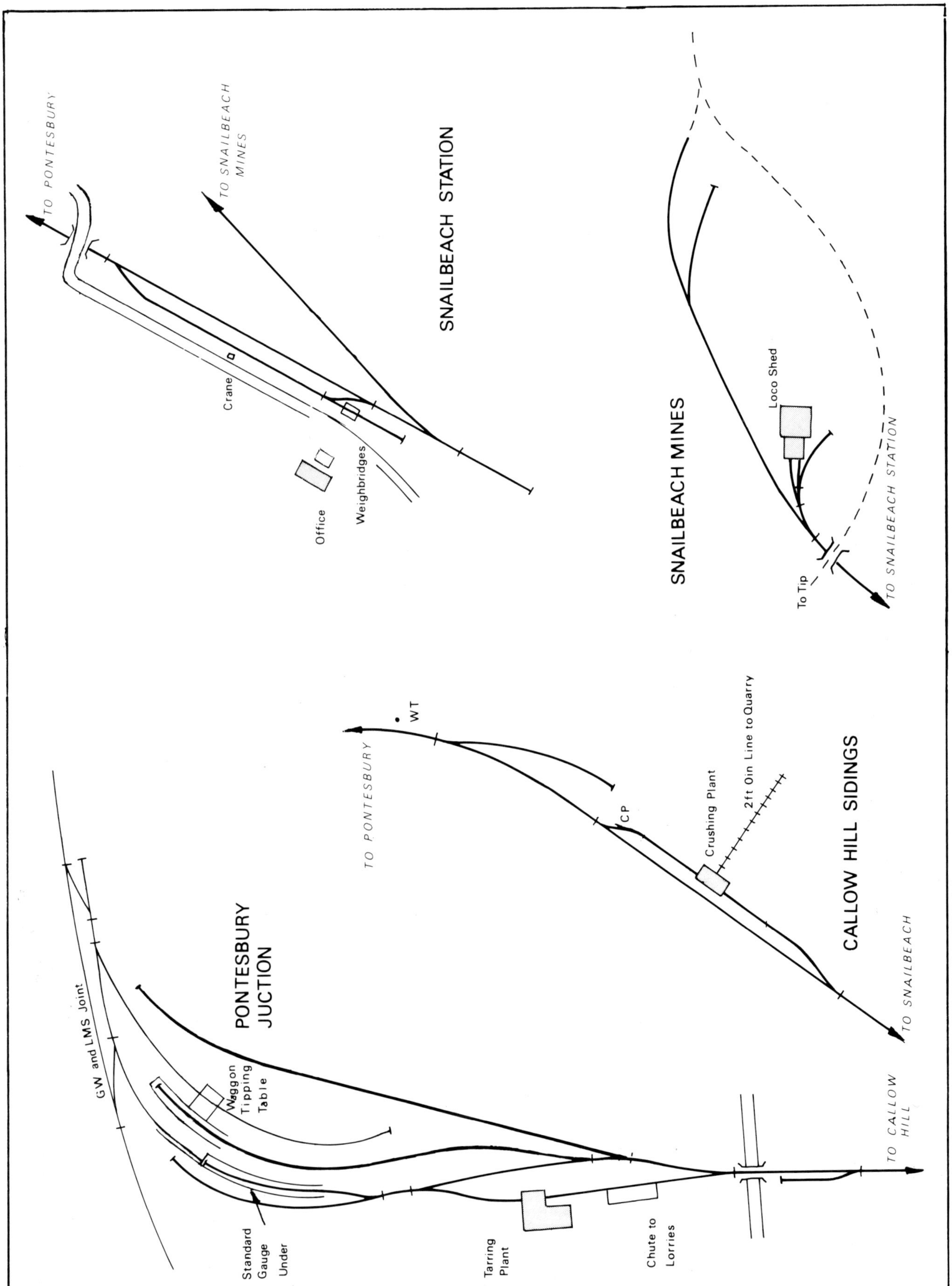
TO PONTESBURY
TO SNAILBEACH MINES
SNAILBEACH STATION
Crane
Office
Weighbridges
SNAILBEACH MINES
Loco Shed
To Tip
TO SNAILBEACH STATION
TO PONTESBURY
WT
CP
Crushing Plant
2ft 0in Line to Quarry
CALLOW HILL SIDINGS
TO SNAILBEACH
PONTESBURY JUCTION
GW and LMS Joint
Waggon Tipping Table
Standard Gauge Under
Tarring Plant
Chute to Lorries
TO CALLOW HILL

227. No 2 and train of hopper wagons leaving the sidings at Pontesbury, 11 June 1943.

228. The station at Snailbeach showing the loading bay and yard crane c. 1926. At the end of the loop are some of the ex-WDLR wagons bought by Stephens in 1923.

229. Baldwin 4-6-0T No. 3 at Snailbeach Mines shed c. 1948/9.

230. Snailbeach locomotive shed with Kerr Stuart 0-4-2T No. 2, Baldwin 4-6-0T No. 4 and the remains of Bagnall 0-6-0T No. 1, *Dennis*, c. 1926.

231. Snailbeach Mines and locomotive shed with No. 2 and the other Baldwin 4-6-0T No. 3 in October 1931. The remains of No. 1 are now cut-off on their siding by a heap of earth.

232. Ten years later (11 June 1943) at Snailbeach Mines all three locomotives are present with No. 2 in steam.

15

The Festinog and Welsh Highland Light Railways

The Festiniog and Welsh Highland Railways together with a large number of quarry lines formed an extensive network of 1 ft 11.5 in gauge railways centred on Portmadoc (or Porthmadog as it is now known). The Festiniog opened on 20th April 1836 as a horse tramway and its subsequent history as the pioneer narrow gauge steam railway is both too well known and beyond the scope of this study to be discussed further. The Welsh Highland originally opened as the North Wales Narrow Gauge Railway on 21 May 1877 and was eventually, as the Welsh Highland, connected to Portmadoc on 1 June 1923. Stephens was appointed Civil Engineer and Locomotive Superintendent on 1 April 1923, by which time both lines were under common management. The Welsh Highland closed to passenger traffic on 26 September 1936 and to all traffic on 31 May 1937. The Festiniog kept going until 1946 and is now thriving as a tourist line.

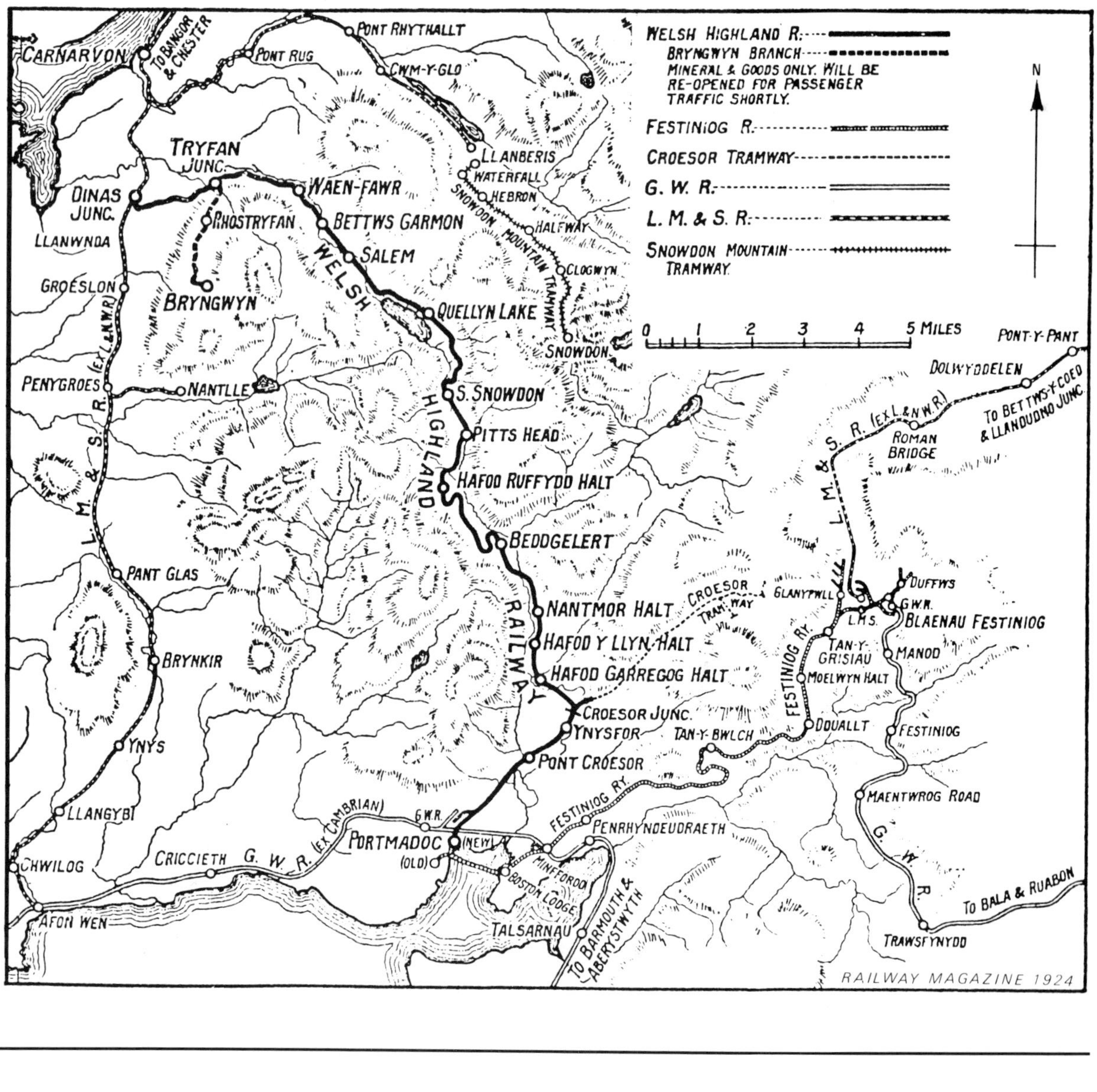

233. A Festiniog train consisting of 0-4-4-0 double Fairlie locomotive *Taliesin* on a train of assorted Festiniog bogie and four-wheeled stock having arrived in the main platform at Portmadoc Harbour in the late 1920's.

234. The locomotive shed at Boston Lodge on 3 June 1932, looking towards Blaenau Festiniog.

235. A Fairlie simmers at Minffordd with a passenger train for Portmadoc c. 1930.

236. *Taliesin* c. 1935 at Penhryn with rolling stock in the light brown (pink or yellow ?) livery of the late 1920's.

237. A double Fairlie and short train under the slate tips at Blaenau Festiniog LNWR station about the turn of the century.

238. A Welsh Highland train waiting to depart from Portmadoc Harbour (FR) to Portmadoc New (WHR) c. 1930. FR England 0-4-0ST+T, *Welsh Pony*, on a train of Welsh Highland stock stands on the connecting line which will take the train through the streets of Portmadoc and across the GWR on the level.

239. Another, identical, Welsh Highland train at Portmadoc New c. 1930.

240. Bryn-y-felin Bridge, near the head of the Aberglaslyn Pass, from the west bank of the Afon Glaslyn in 1923.

241. *Welsh Pony* at Beddgelert with a train from Portmadoc in the early 1930's.

242. The station building at Beddgelert c. 1930. If you look carefully you can just see the station mistress in her Welsh costume under the station shelter.

243. Hunslet 2-6-2T, *Russell*, at Beddgelert with a train from Portmadoc in the late 1920's.

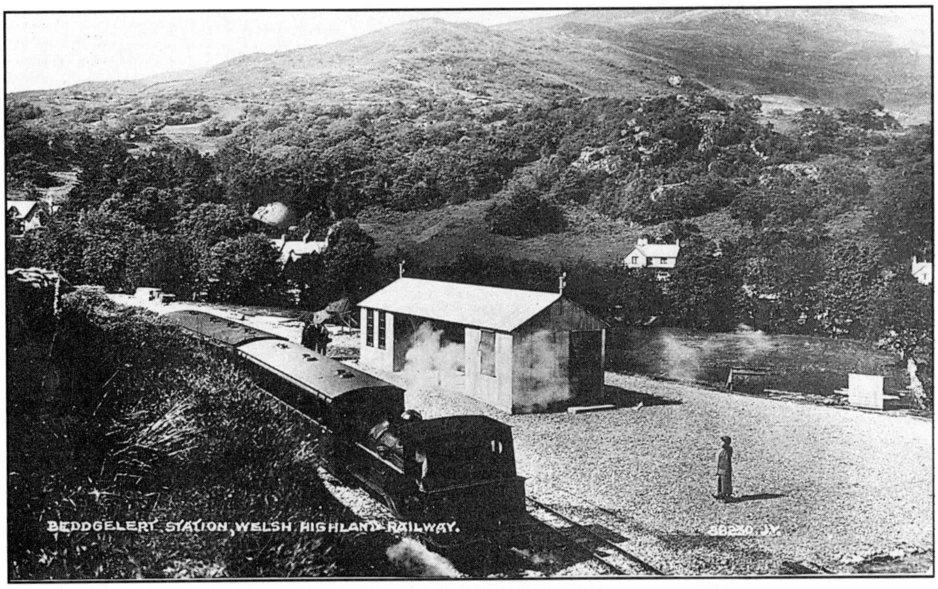

244. Ex-NWNGR Spooner designed single Fairlie 0-6-4T, Moel Tryfan, at Beddgelert c. 1923.

245. Nantmor Station with a train composed wholly of Festiniog stock heading north in 1923.

246. A general view of Dinas Junction locomotive shed showing a WHR bogie all-third to the left and the bunker of the Baldwin 4-6-0T in the right hand road of the shed, 31 August 1926.

247. Another view of the shed area at Dinas Junction taken from the over bridge in 1934.

16

The Ashover Light Railway

This 7.25 mile long 60 cm gauge line was opened between Ashover and Clay Cross on 6 April 1925. It closed to passengers on 30 September 1936 and totally on 31 March 1950. Stephens was the Engineer.

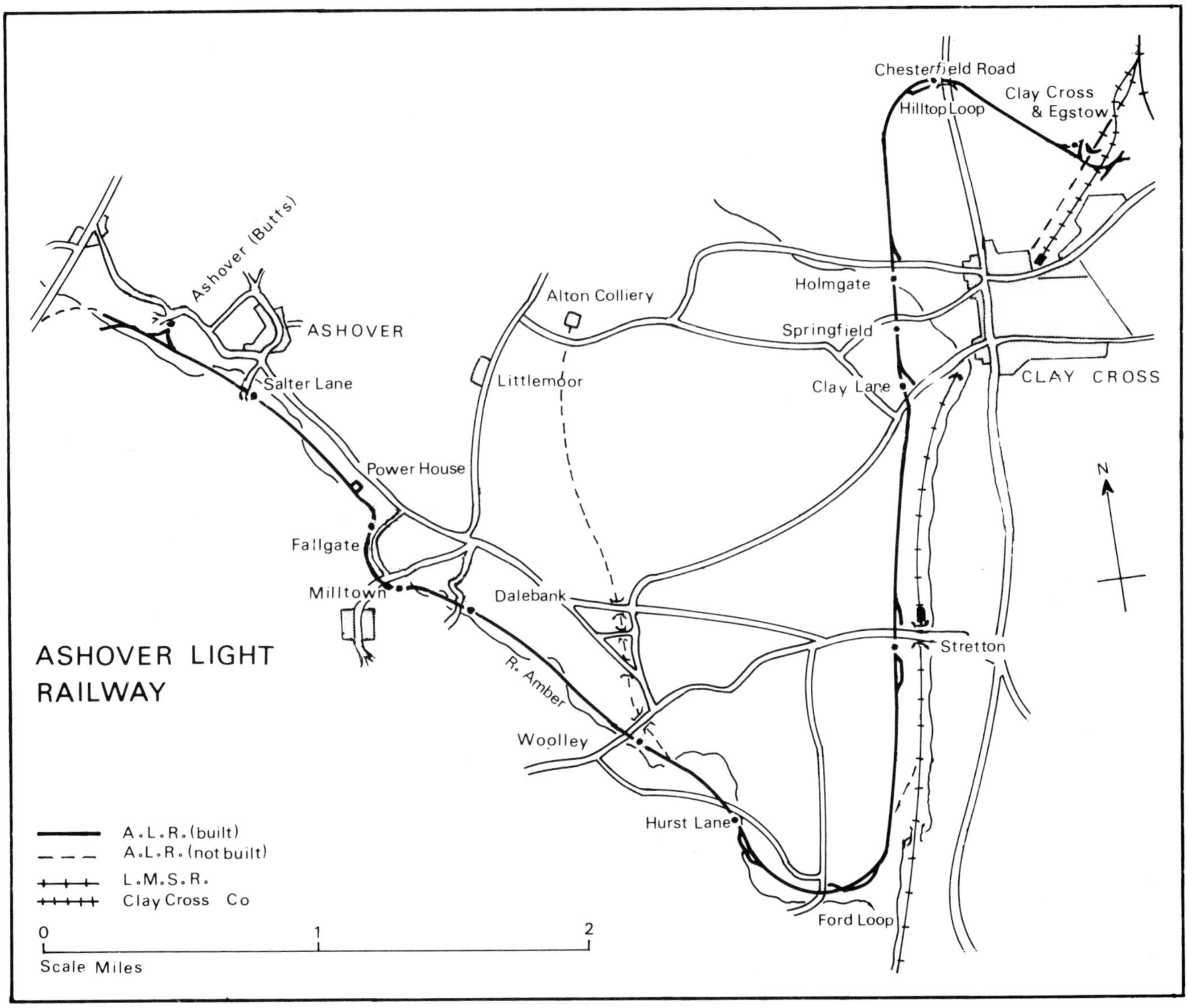

248. Clay Cross Station c. 1930 showing the waiting shed, loop and three of the Neverstop Railway carriages from the British Empire Exhibition at Wembley.

249. Baldwin 4-6-0T, *Hummy*, at Clay Cross with a one carriage off-season train c. 1925.

250. *Peggy* at Clay Cross with two of the Gloucester Carriage & Wagon coaches c. 1927.

251. Baldwin 4-6-0T, *Peggy*, at Chesterfield Road on a mixed train for Ashover c 1935.

252. *Hummy* at Hill Top Loop with a train for Clay Cross. Note the makeshift toast rack carriage with canvas roof converted from one of the bogie open wagons. Two of these were used in 1925 when the four Gloucester coaches proved insufficient. They were replaced by the first of the Neverstop Railway carriages to enter service in 1926.

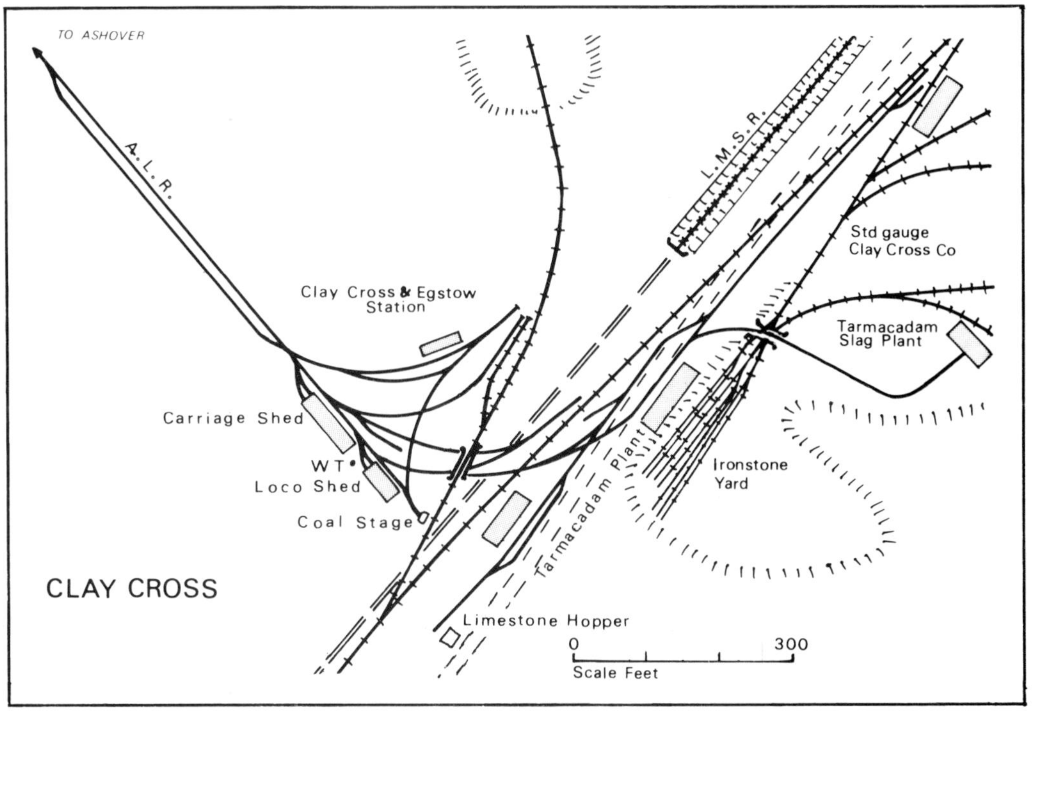

253. *Bridget* at Springfield with a long line of bogie wagons c. 1945.

254. *Hummy* leaving Stretton on 26 August 1947 with a loaded stone train for Clay Cross.

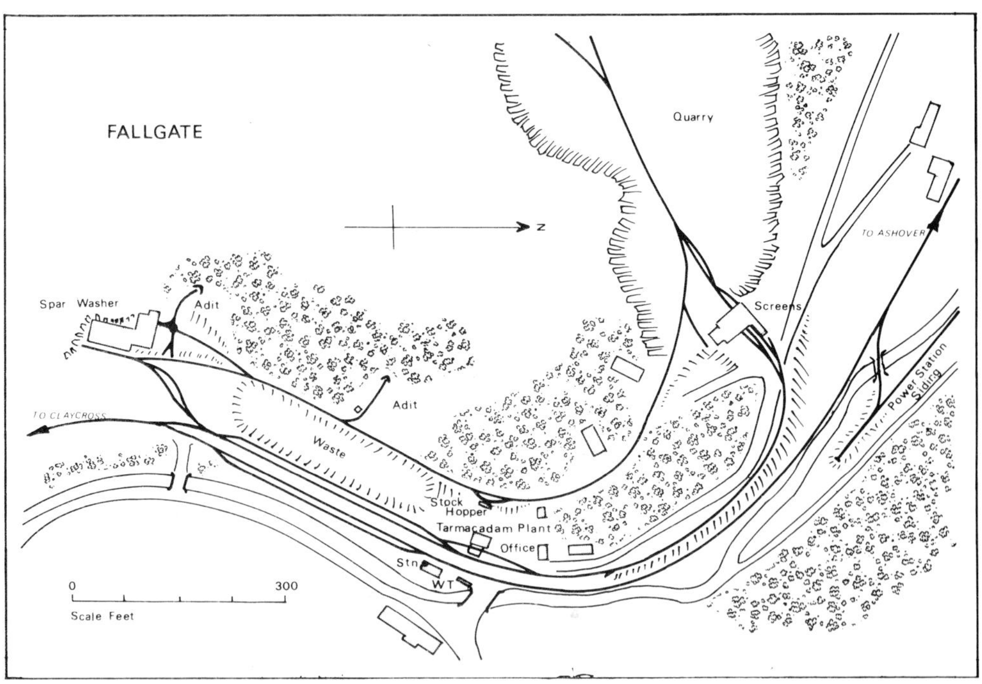

255. Fallgate yard looking towards Ashover in 1946.

256. *Peggy* taking water from the cylindrical tank in Fallgate Yard c. 1945.

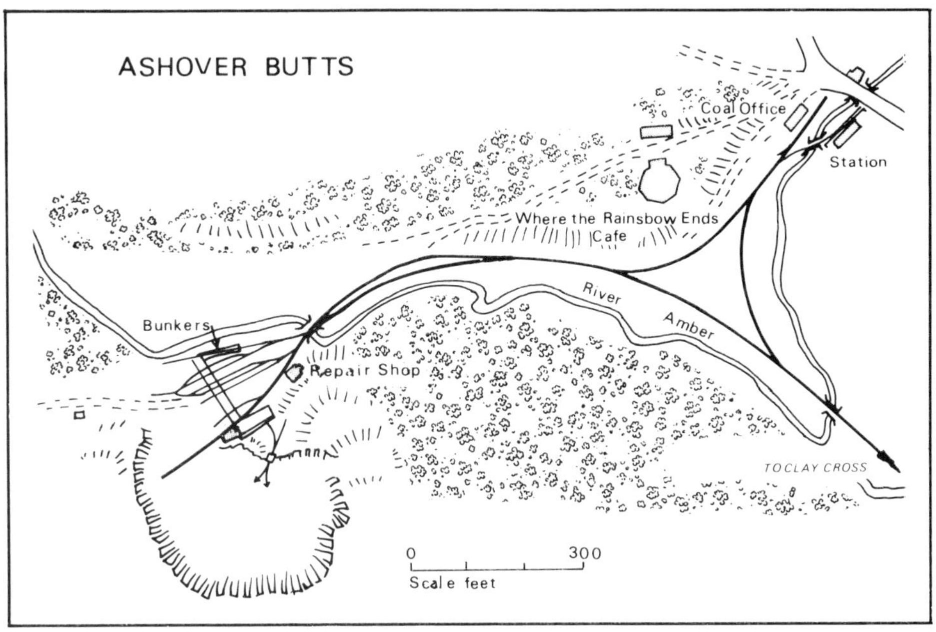

257. Ashover Butts Station c. 1935.

258. Ashover Butts Station c. 1926. The sign on the ice cream kiosk behind the station reads "Stop try our ales".

259. The triangle at Ashover Butts with the *Where The Rainbow Ends* cafe in the background. This was opened by the Railway in 1926 and soon became popular for both meals and summer evening dancing.

17

The North Devon and Cornwall Junction Railway

This 20.5 mile standard gauge line from Torrington to Halwill Junction was partly built on the track bed of the 3 ft gauge Torrington & Marland Railway and opened on 27 July 1925. It was closed to all traffic from Halwill to Meeth and to passengers throughout on 1 March 1965. The last train ran on 6 Nov 1982. Stephens was the Engineer.

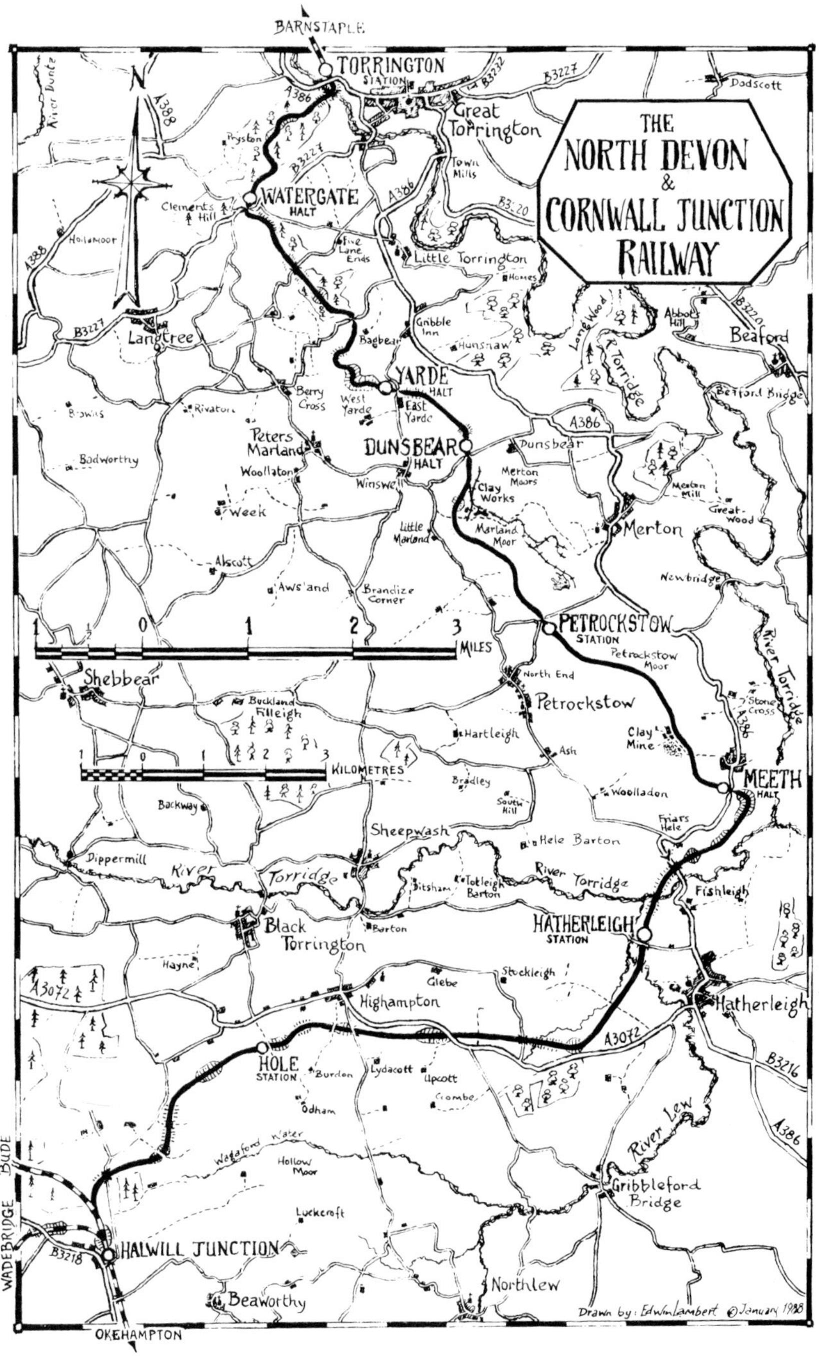

261. Torrington Viaduct on 9 April 1977 showing the remains of the piers of the original 1881 J. B. Fell designed viaduct of the Torrington & Marland Railway still visible in the river.

260. Torrington Viaduct on 9 April 1977, looking south. This was the largest structure designed and built by Stephens.

262. In addition to the Torrington Viaduct, Stephens replaced all of the old wooden structures of the Torrington & Marland Railway as typified by the road bridge at Badslake, about 1 mile south of Watergate, photographed on 9 April 1977.

263. Rural splendour at Watergate Halt looking towards Torrington c. 1960. This platform was Southern Railway built several years after the line opened.

264. Watergate Halt with Ivatt 2-6-2T No. 41290 on a single Bullied brake third in 1964.

265. Yarde Halt, another Southern built platform, with the tail end of a mixed train departing for Torrington in the summer of 1962.

266. Continuing its progress on that same day in 1964, No. 41290 is now seen waiting at Dunsbear Halt.

267. Another view of Dunsbear Halt, probably c. 1960, showing the platform and shelter built, unusually for Stephens, of stone. Note the Stephens style wooden bench.

268. No. 41290 again, now at Petrockstow continuing its journey south (on that same day in 1964?)

269. Petrockstow station on 9th July 1962. Ivatt 2-6-2T No. 41297 waits at the platform with a mixed train for Torrington.

270. Meeth Station showing the original stone built ND&CJLR shelter surviving 12 years after closure on 9 April 1977.

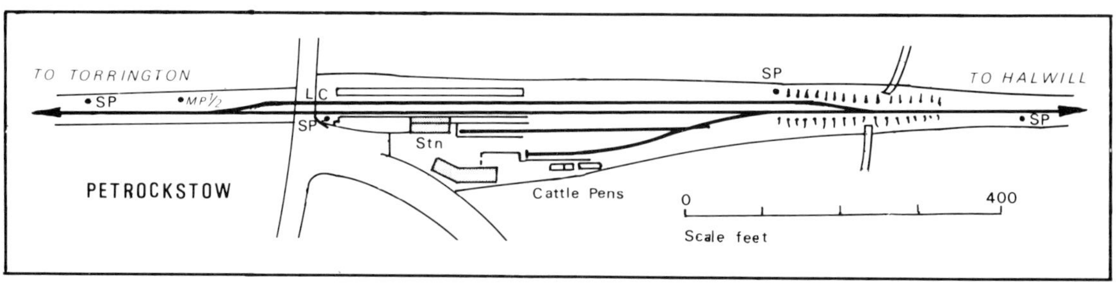

271. One of the ex-LBSCR E1R Class 0-6-2T's rebuilt from an E1 Class 0-6-0T by R. E. L. Maunsell in the early 1920's seen here at Hole Station, c. 1936.

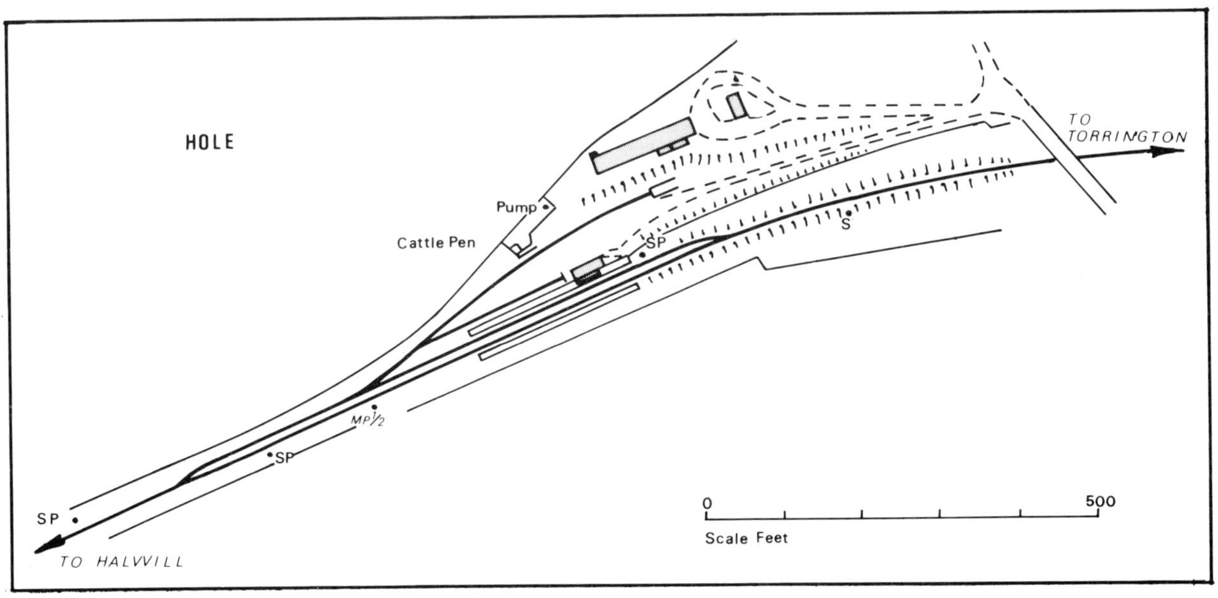

Kent & East Sussex Light Railway. Mixed train from Headcorn to Tenterden. circa 1953, with British Railways A1X Class locomotive No. 32655.
Edwin Lambert - August 1981.

18

Tailpiece

EVENING ALONG THE ROTHER VALLEY

Twenty minutes late, the down Hastings stopping train, headed by a Schools Class 4-4-0, made its way over the smooth permanent way that snaked southward from Tonbridge to the coast.

The gentle click clack of those wheels on the joints south of Tunbridge Wells Central, as we headed out into the last light of that winter's day, had a quiet reassuring and calming effect as I nodded off in a peaceful slumber. It seemed all too soon when the train eased into the platform at Etchingham and, as if a moment later, the train again slowed as we approached Robertsbridge, passing the yard and outer home signal, with its green spectacle giving the right of way. All of a sudden we were in the platform, with its dimly lit electric lamps casting a shadowy light across the tarmac surface. Within a minute or so the main line train had departed into the night taking with it that warmth and reassurance that only a corridor train has, the sound of the Schools Class echoing down the grade as the locomotive ran south hauling its string of yellow lit Maunsell corridor stock.

The station building was well lit and looked very inviting on that cold winter's night as I walked up the empty platform. I entered the main hall and sat waiting for my connection to Northiam. After about twenty minutes the sound of a train could be discerned in the distance and, presently, it was announced that the connection from Tenterden Town had arrived. I gathered my bags and walked up the platform towards my first sight of the gypsy train that was to transport me through the black of the night.

The little Emmett-like Terrier locomotive ran round its solitary ex-LSWR bogie corridor brake third and we were soon ready for the road. A young fireman changed the headlamps on the locomotive as it simmered in the bleak, dimly lit station platform.

After a little while a guard gave a short blast on his whistle and we were soon on the move across the yard and then, veering off sharply to the right away from the main line and towards Salehurst, the little Brighton tank descended the short section between Robertsbridge Station and Hodson's Mill. A pillar of white steam rose skywards above the wintry bare trees that lined the embankment. We crossed the road and continued along the Rother Valley, the train settling down to a gentle pace until we reached an empty platform in the middle of a field; Salehurst Halt served two houses, a pub and a church. Nobody left or joined the train and we were soon on our way.

It was starting to drizzle as we passed through the hop fields near Junction Road Halt; the lights of the bogie saloon danced with the fleeting shadows of the frames for the hop plants in the adjoining fields. We briefly halted at Junction Road and then proceeded to Bodiam along the banks of the River Rother past the Guinness hop fields and across a number of culverts until in the distance the white bleached walls of Bodiam Castle loomed up, the moonlight glinting across the turrets and picking out the battlements through a break in the black night clouds. We ran into Bodiam Station, where a group waited on the platform to join the train. The Terrier braked gently and waited while the passengers boarded and the Station Master loaded parcels into the brake, a wisp of white vapour slowly drifting into the night air.

A whistle from the locomotive suddenly broke the silence and we were, in a moment, in motion again rolling through the dark fields of hops, past ghostly sihouettes of wintry trees with only the occasional telegraph pole for company. The wires of the boundary fence played tricks with the light of the locomotive's fire as we plodded along the valley towards Northiam.

The beat of the little Terrier made a brave sound in the stillness of the cold night - the chatter on the bullhead rails and the bounce on the rail joints. All too soon we approached the yard at Northiam over the points and into the empty platform. I departed my compartment and walked up the platform standing only for a moment to watch the little train depart away into the night, through the hedge-rows and empty fields, on its way to Tenterden. It had been quite an evening.

272. *TAILPIECE:* K&ESR No. 8, *Hesperus,* in store at Rolvenden c. 1935.

Appendix

KENT & EAST SUSSEX RAILWAY BUILDINGS
DRAWINGS AND HISTORICAL NOTES

by Leslie Darbyshire

INTRODUCTION

This Appendix provides plans of track layouts and buildings at a selection of stations on the Kent and East Sussex Railway. They were produced from measurements taken on the site over the period 1955 to the present day, augmented by information from 25-inch Ordnance Survey maps dated 1908-9, and by many photographs. These show that there were changes over the years and some of these are indicated on the drawings or in these Notes.

BODIAM STATION

This was the first station on the Rother Valley Railway and the original 1900 station building survives today, almost unchanged. There were two sidings on the Up side, one of which ran alongside a timber-faced loading dock, probably for cattle. The level crossing was gated (possibly with only one gate) and a 2-arm signal was sited on the platform for indicating when trains were to stop at the station.

About 1910 a siding was added on the Down side to enable passenger and freight trains to pass; two passenger trains were not allowed to do so unless one of them was the steam railcar. The siding was never made into a loop. At around this time the 2-arm signal was removed as were the crossing gates.

The original R.V.R. track was of light flat-bottomed rail spiked directly to the sleepers. The main line was relaid with bullhead track at an early date and the Down siding somewhat later, but the two Up sidings retained their R.V.R. track until closure.

The main station building stood on a low concrete plinth and was constructed of timber framing clad in the standard material of most Colonel Stephens stations, corrugated iron. It was slightly asymmetrical and had a roofless 'gents' at the Tenterden end and a lean-to store at the other. The platform was faced with concrete with a brick edging.

Even today, Bodiam station in its isolated setting retains much of the charm and character of the Colonel Stephens railways and it is to be hoped that the eventual extension of the preserved K.E.S.R. to it does not destroy this; it would be a tragedy if it did.

NORTHIAM STATION

This was the second station on the R.V.R. and had the only passing loop. Two platforms were provided, together with a building generally similar to Bodiam but considerably larger and with the 'gents' and store both at the same end. Although two doors were provided on the 'road' side of the building and there was a considerable difference in level between the platform and the ground, curiously only one door had steps for access.

Two sidings on the Up side were provided, one with extensive loading facilities for livestock and goods, including an end-loading carriage dock. The other siding also had a small side-

273. Staff bungalow at Northiam.

and end-loading dock just behind the platform; this is shown on an early R.V.R. plan and is partly visible in a 1908 photograph. It is not clear how access was obtained to this dock as there was no gate from the platform and it was well above ground level.

The level crossing was again originally gated, photographic evidence suggesting only one gate. This gate was removed at an unknown date and cattle grids substituted. Also at an unknown date, two staff bungalows were built on the Southern boundary of the site.

As at Bodiam, some flat-bottomed track survived in the sidings until closure although by this time the second platform had gone. In other respects the station was and is largely unchanged, though a coal depot now occupies the Up side of the site.

The Tenterden Railway Company has plans to extend the preserved K.E.S.R. to Northiam by 1990, and this will change the site's appearance considerably as they intend to reinstate the second platform (a good move) and install a footbridge (a bad one, out of keeping with the line's character).

ROLVENDEN STATION

This was the original R.V.R. terminus and was named 'Tenterden' until the line was extended to the present Tenterden Town station in 1903. It had a building of similar size to Northiam but more symmetrical. The usual two Up sidings were provided, with a run-round loop and 2-road locomotive shed on the Down side. The extension to Tenterden across a gated crossing over the present A28 road removed much of the station's traffic, but it remained the line's operating centre as far as locomotives and rolling stock were concerned.

By 1908, the run-round loop had been considerably lengthened and two sidings built on the Down side; one of these was covered by a shed described as a paint shop on the early plan. On the platform a second building had been added; this was used as a printing works for posters etc. for many of the Colonel's lines. At a later date three more Down sidings were added which were used largely for the storage of redundant stock, of which there was a great deal. Behind the furthest of the sidings, one or more large haystacks were generally to be found during the Summer; these were produced from grass cut from embankments and cuttings along the line.

The station platform was originally timber-faced but was later rebuilt in brick and acquired a portion raised about a foot above the general level for the loading and unloading of goods. There may have been a loading dock behind the platform, as at Northiam, as there were double gates in the fence in a suitable position. Again as at Northiam, only one of the road-side doors to the station building was fitted with access steps.

The track in the sidings remained flat-bottomed until removal along with all buildings in the mid-1950s. Unfortunately, little remained to help the preservation society, even the platform face having been demolished.

ROLVENDEN LOCOMOTIVE SHED

A 2-road shed was opened in 1900 at the opening of the line. It was about 50 feet long with a curved corrugated iron roof. Before 1908 the shed had been lengthened by about 35 feet and a workshop had been built on the South side. At this time the entire structure including the workshop appeared to be timber-clad and a pitched roof had been fitted. Some time after 1910, the workshop was enlarged and rebuilt in brick, and the shed clad in corrugated iron.

In 1900 a water tower stood just outside the shed, between the shed roads and the running line. By 1908 that tower was on the side of the shed itself. It is possibly that it was in the same position as before and had simply been engulfed by the shed extension. A brick-built coaling platform stood alongside the tower. A second water tower was later built next to the first, with yet another at the end of the platform. This latter tower originally stood on brick pillars, but was eventually replaced by a new tank on reinforced concrete supports.

The shed was lit by roof lights and side windows and was ventilated by roof ventilators. Later the latter were supplanted by louvred gable-ends which were sheeted over in the 1930s. Some of the side windows and the roof lights were also sheeted over towards the end of the building's life which came in 1955.

ROLVENDEN PAINT SHOP

This building was, as stated earlier, described as a paint shop on a K.E.S.R. plan, but may have been built to house the 1905 steam railcar; certainly the evidence of photographs suggests that most painting, at least in the 1930s, took place outdoors. The building was entirely timber built, and at some time after 1908 acquired a lean-to extension as shown on the drawings. The main building had doors at both ends, and the imprints of hinges in the uprights of the lean-to suggest that it was at one time similarly fitted.

Outside the building were two sleeper-built walkways used for carriage servicing and the painting operations mentioned above.

The whole structure lasted until Nationalisation but had been demolished by 1955. Fortunately the foundations were still in situ and could be measured.

TENTERDEN TOWN STATION

When opened in 1903, Tenterden Town had a single platform facing a run-round loop. On the platform was a simple timber building clad in the usual corrugated iron. When the extension to Headcorn was built in 1905, a much more imposing station (by light railway standards anyway) was provided. An island platform was built on the opposite side of the loop with a terminal road on its far side, facing Headcorn. The old building was moved to Headcorn, where it lasted until closure, and a new,

274. Tenterden yard crane in 1955.

larger building provided. This was quite unlike any other on the line and was built in brick which was pierced by arches with matchboarded timber inserts containing the doors and windows. Presumably this odd method of construction was to save cost.

In addition to the terminal road, a siding was provided on the Down side and four more on the Up side. One of the latter served a small cattle dock (the front wall of which still exists) with direct access to the market which was once held just South of the station. The next two sidings at one time served a side- and end-loading platform, but this was later removed and a hand operated 2-ton crane installed.

The crane still exists, and is the subject of a separate drawing. It has a timber jib and rotates around a timber column set in a square brick-built platform. Although it appears to be very old, it was not installed before about 1910; it may well have been second-hand.

Some time after the opening of the line, a circular water tank on a cast-iron column was sited at the Rolvenden end of the Down platform. At the opposite end of the same platform was a tall wind pump which presumably fed the tank, and which had originally been situated on the Up side of the line some way North of the station.

At the Rolvenden end of the station was a very curious 3-arm signal which was installed at an early date, probably in 1905. As the drawing shows, it had two arms facing Rolvenden which were Home signals controlling entry to the two main platform lines, and one facing Headcorn serving as Starter for both lines. Judging by its general design and particularly the shape of the finial, it may have been an L.C.D.R. original.

This signal and the adjacent loop points were operated by an open ground frame nearby. The points and signals at the Headcorn end of the site were operated from another ground frame but this time enclosed in a small corrugated iron-clad cabin as shown on the drawing.

Apart from the replacement by brick of the timber insert

at the West end in B.R. days, the station building remains externally intact to this day as the headquarters of the Tenterden Railway Company who, despite the inevitable additions of toilet block, refreshment room, signal box and carriage shed, have nevertheless managed to produce a very attractive station which retains a lot of character.

HIGH HALDEN ROAD STATION

This was the first full station on the Headcorn extension. It had no passing loop, and had the usual two Up sidings. Unlike those on the R.V.R. section, there were never any gates to the crossing.

The building bore a strong resemblance to the R.V.R. stations except that it was clad in creosoted timber weather-boarding rather than corrugated iron, and stood on a brick plinth rather than a concrete one. It was also curiously asymmetric, particularly noticeable on the awning supports which were unevenly spaced.

A 2-arm stop signal, similar to that originally at Bodiam and possibly the same one, was sited on the platform and remained there until closure of the line in 1954.

The station building remains in use as a store on a smallholding.

OTHER STATIONS (not drawn)

Wittersham Road was generally similar to Bodiam, but its building was very unusually, at right angles to the platform. The entire site was cleared by B.R. and the present 'preserved' station is from Borth on the Welsh coast.

Frittenden Road was very similar to High Halden Road; its building was identical and the track layout differed only in its being situated on a gentle curve. The platform was timber-faced whereas High Halden Road had always had a brick-faced platform. The building survives, though it is in a very derelict state.

Biddenden's building was weatherboarded like the other extension stations but was larger, being similar in size to Northiam which it resembled in having passing loop and two platforms, both brick-faced. It survives, somewhat altered, in private hands.

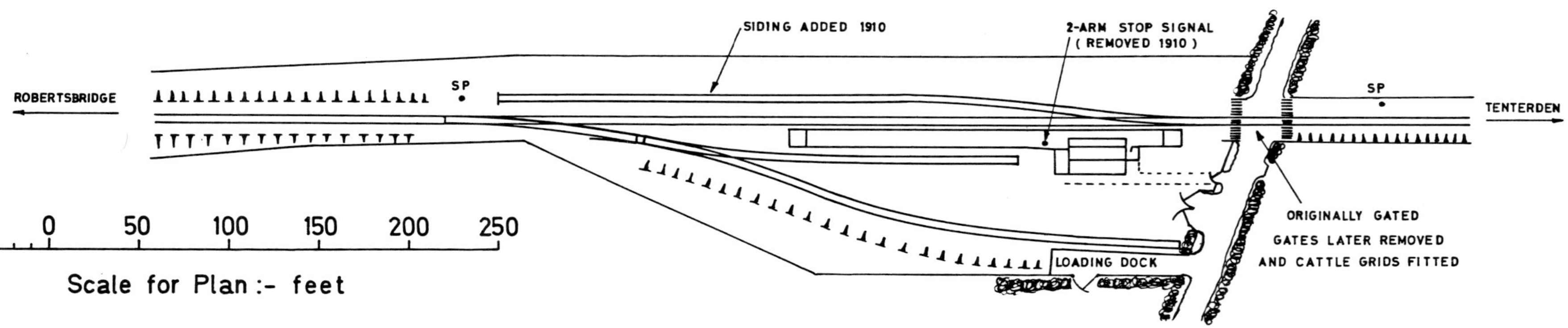

BODIAM STATION

K&E.S.R.

(sheet 1)

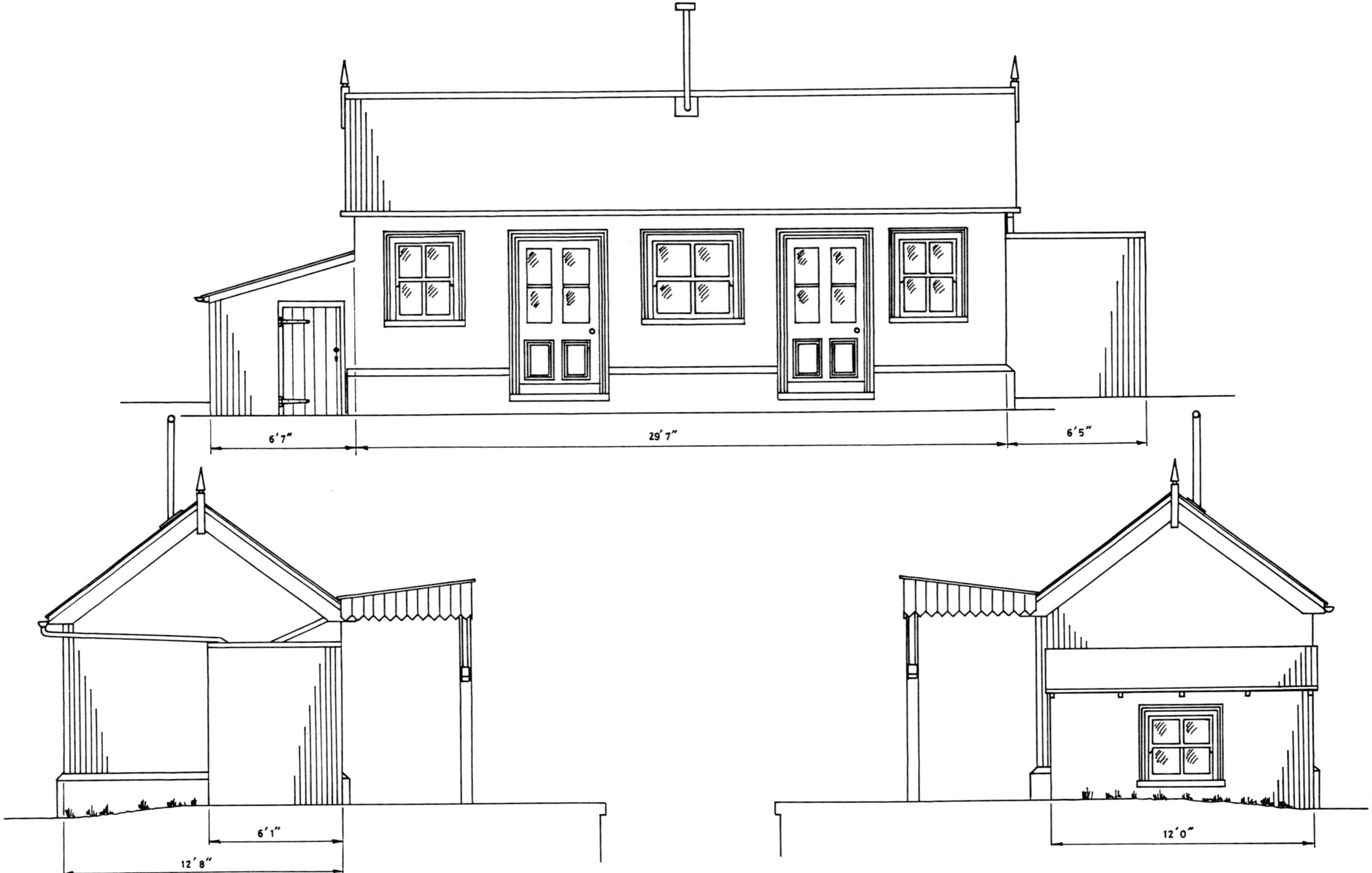

BODIAM STATION
K.&E.S.R.
(sheet 2)

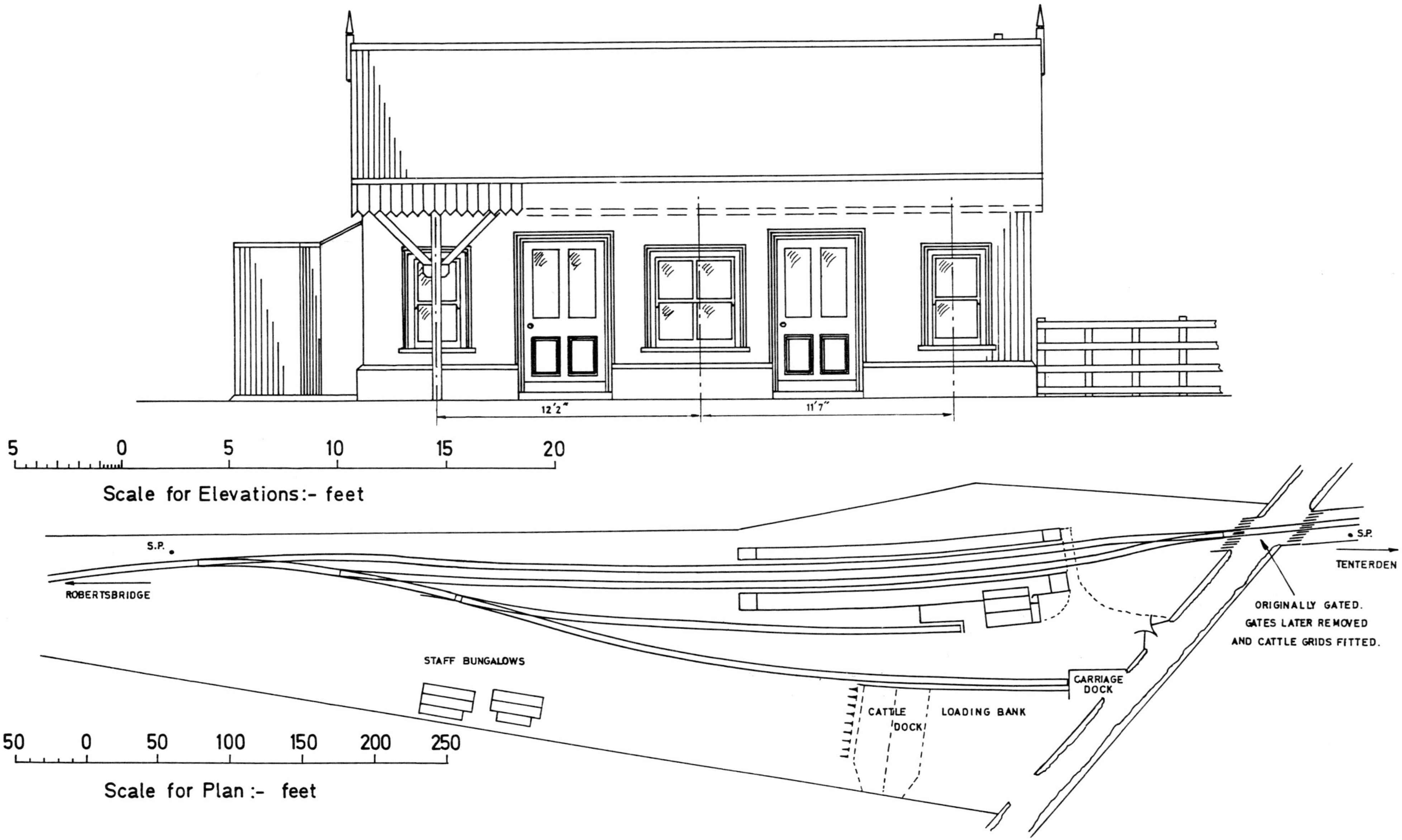

NORTHIAM STATION
K.& E.S.R.
(sheet 1)

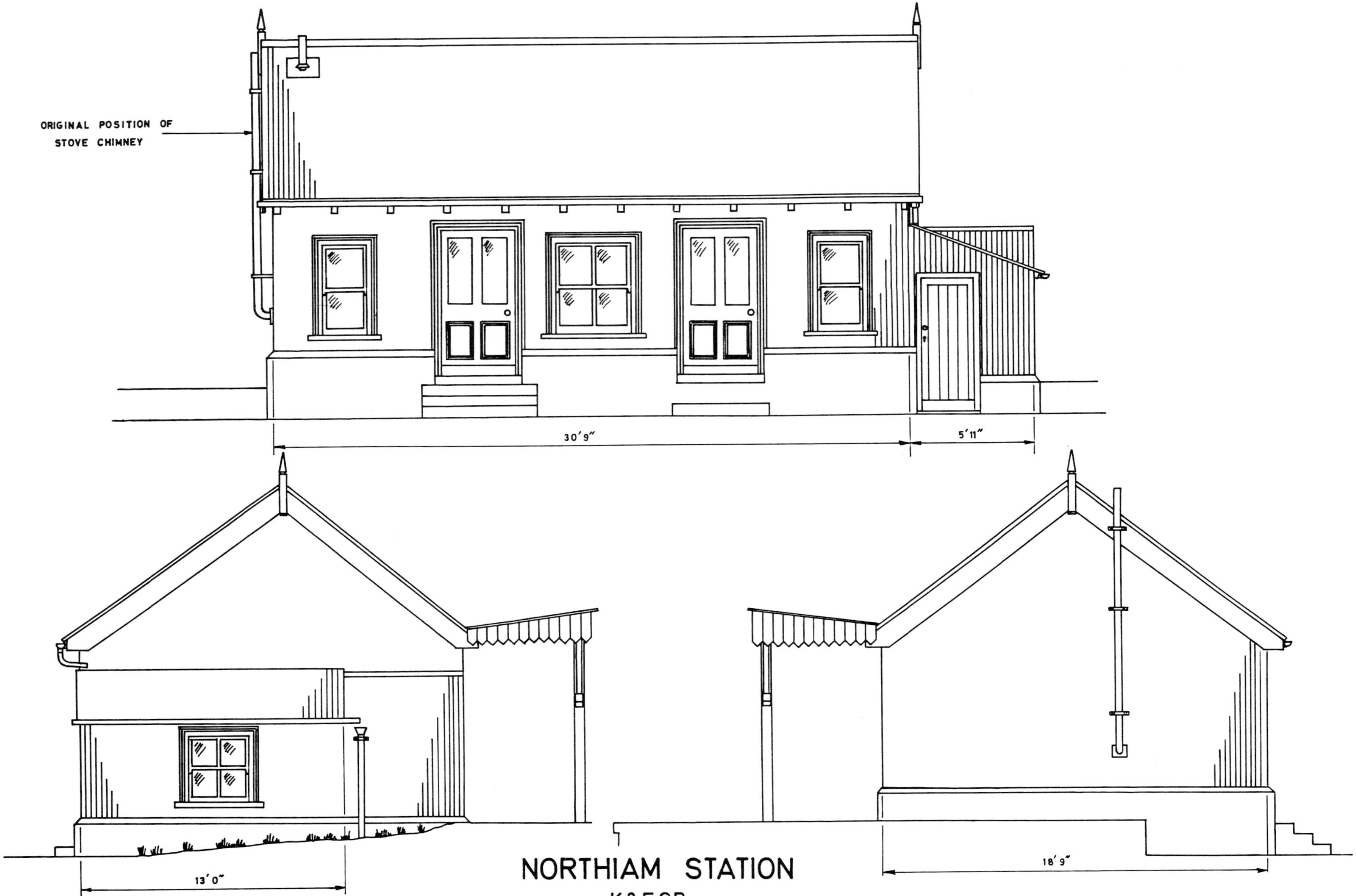

NORTHIAM STATION

K.& E.S.R.

(sheet 2)

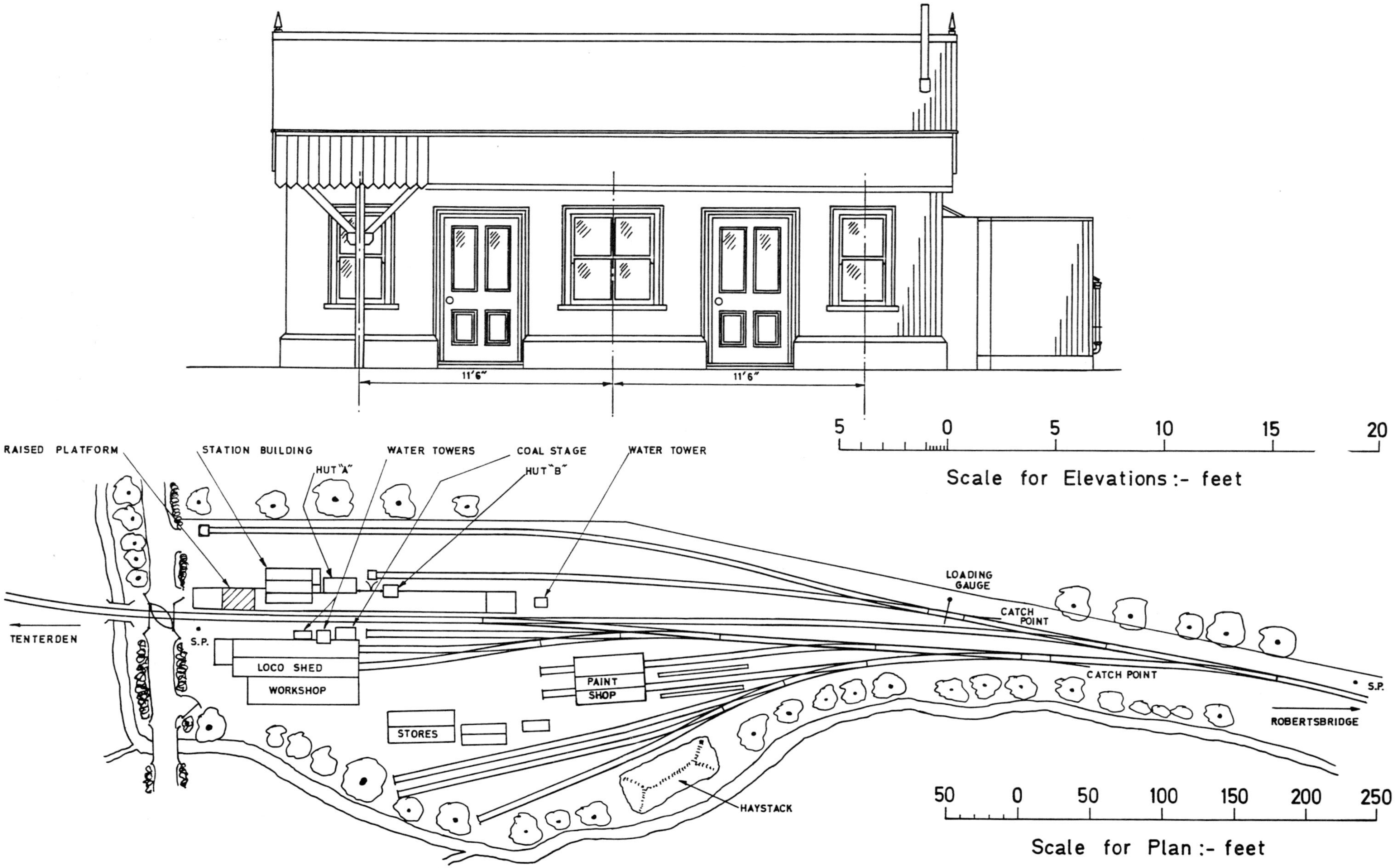

ROLVENDEN STATION

K.&E.S.R.

(sheet 1)

ROLVENDEN STATION

K.&E.S.R.

(sheet 2)

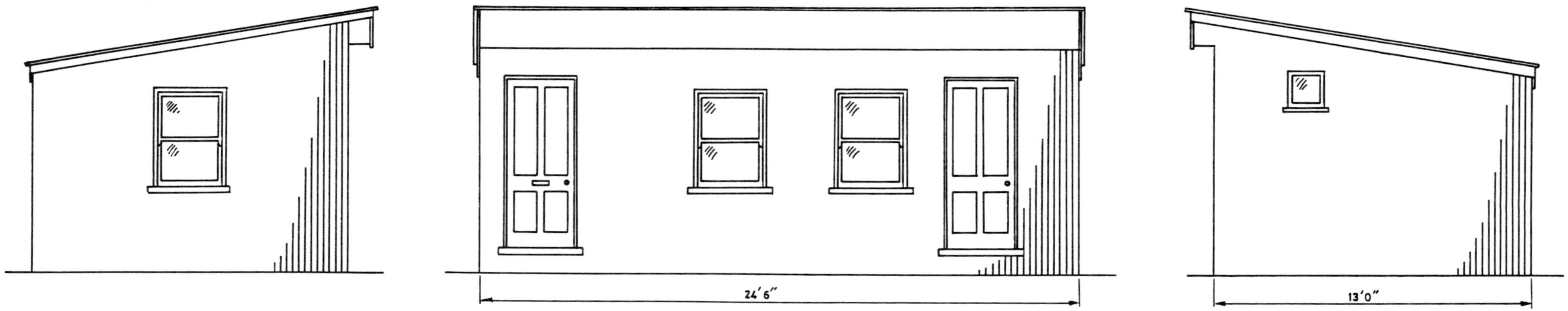

Hut "A" (Printing Shop)

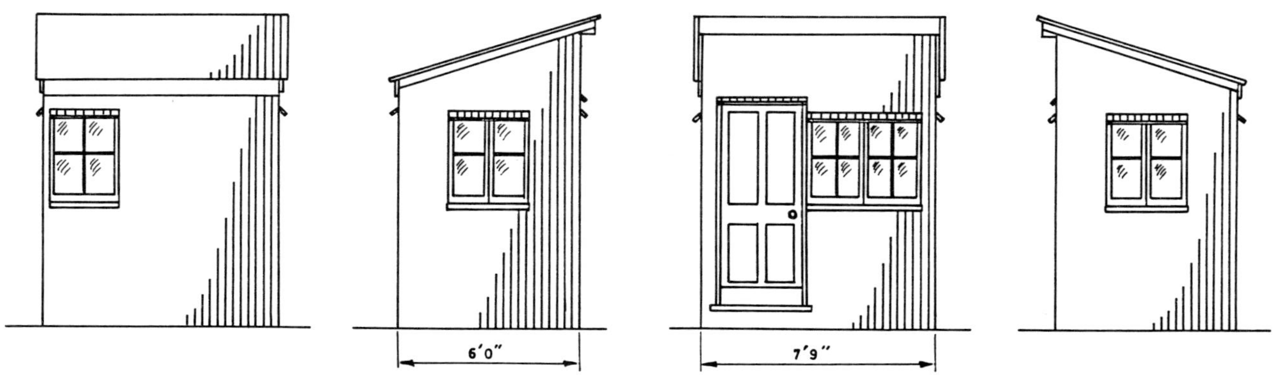

Hut "B"

ROLVENDEN STATION

K.&E.S.R.

(sheet 3)

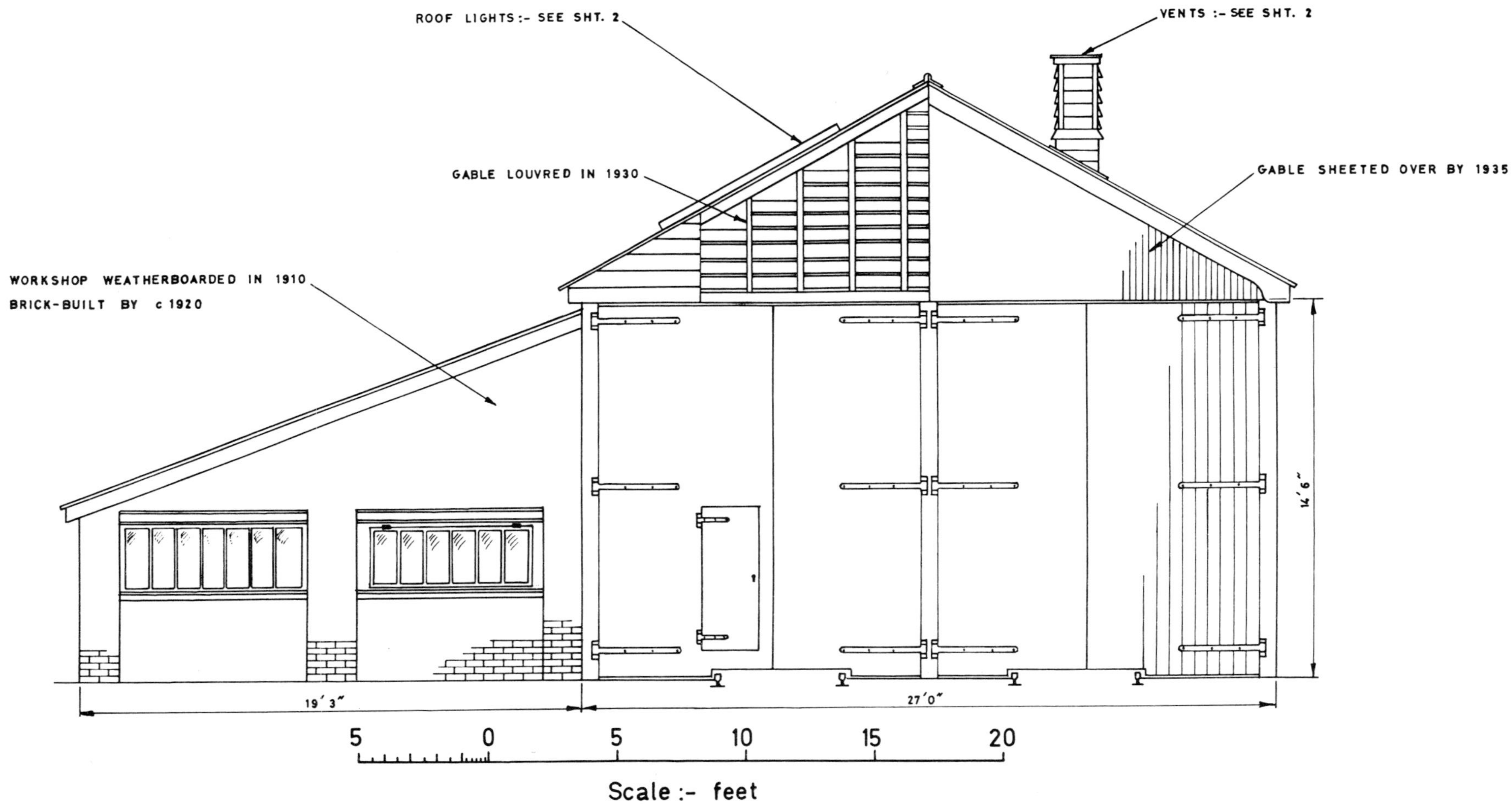

ROLVENDEN LOCO. SHED

K.&E.S.R.

(sheet 1)

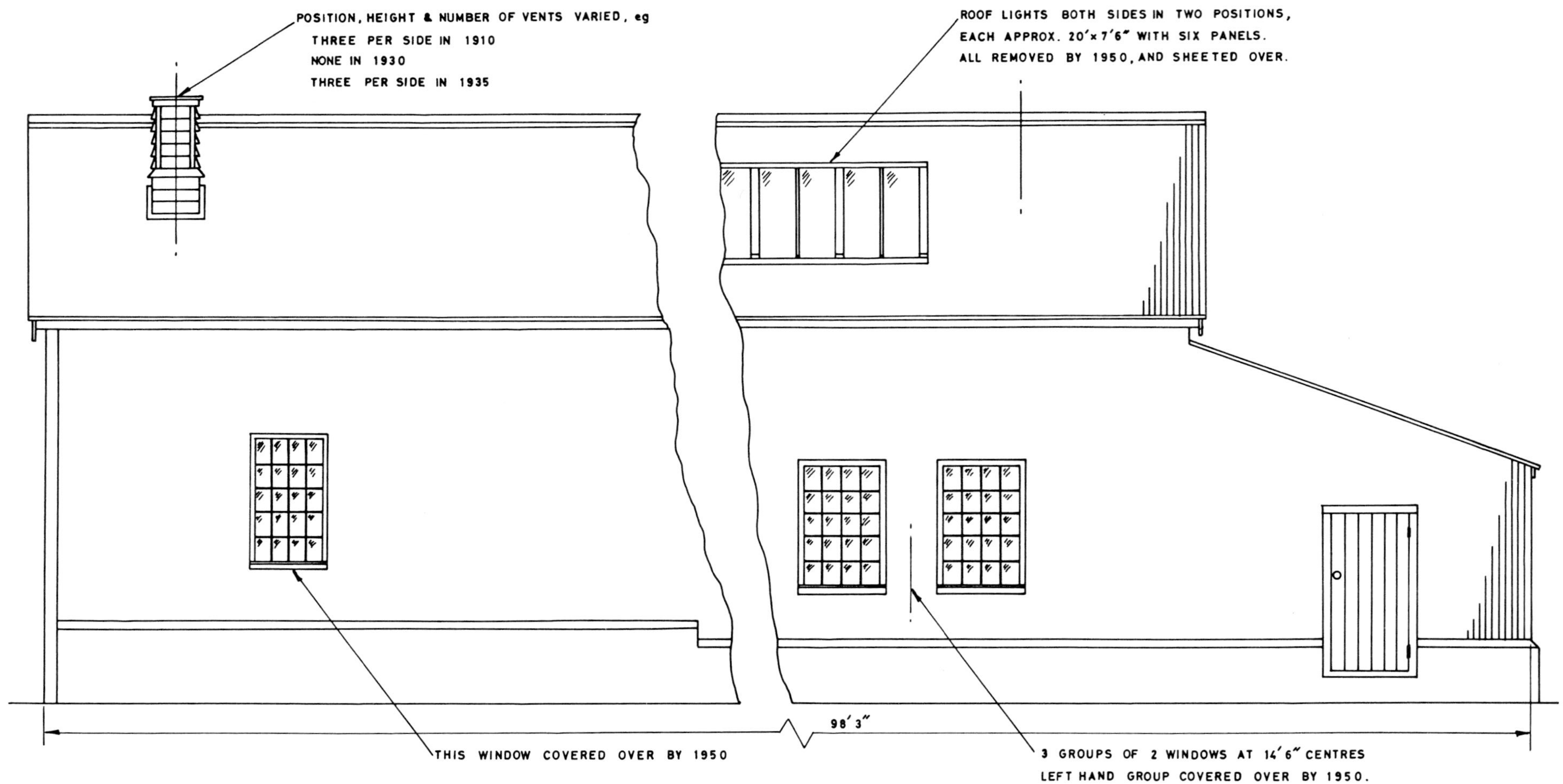

ROLVENDEN LOCO. SHED

K.&E.S.R.

(sheet 2)

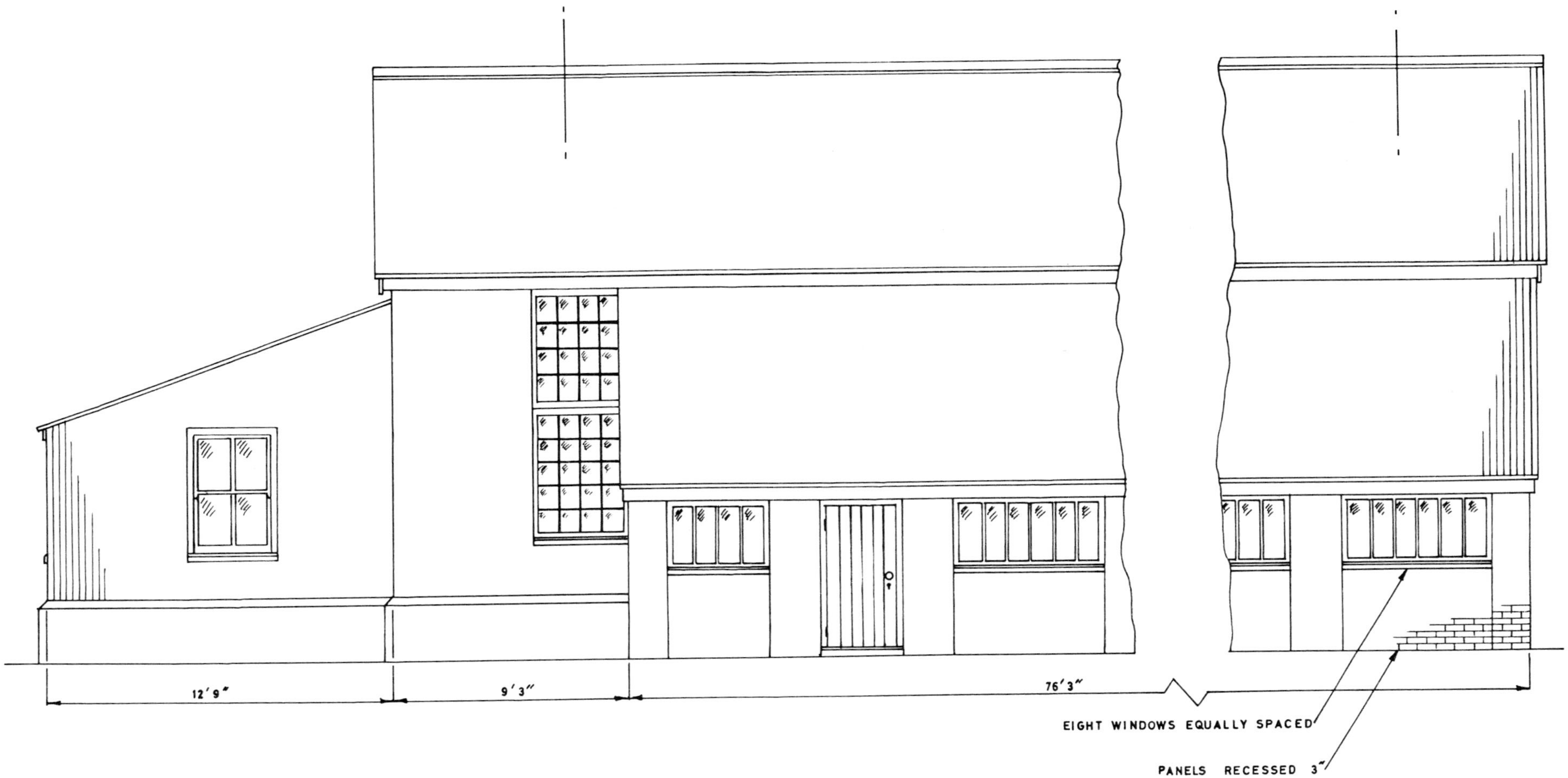

ROLVENDEN LOCO. SHED

K.&E.S.R.

(sheet 3)

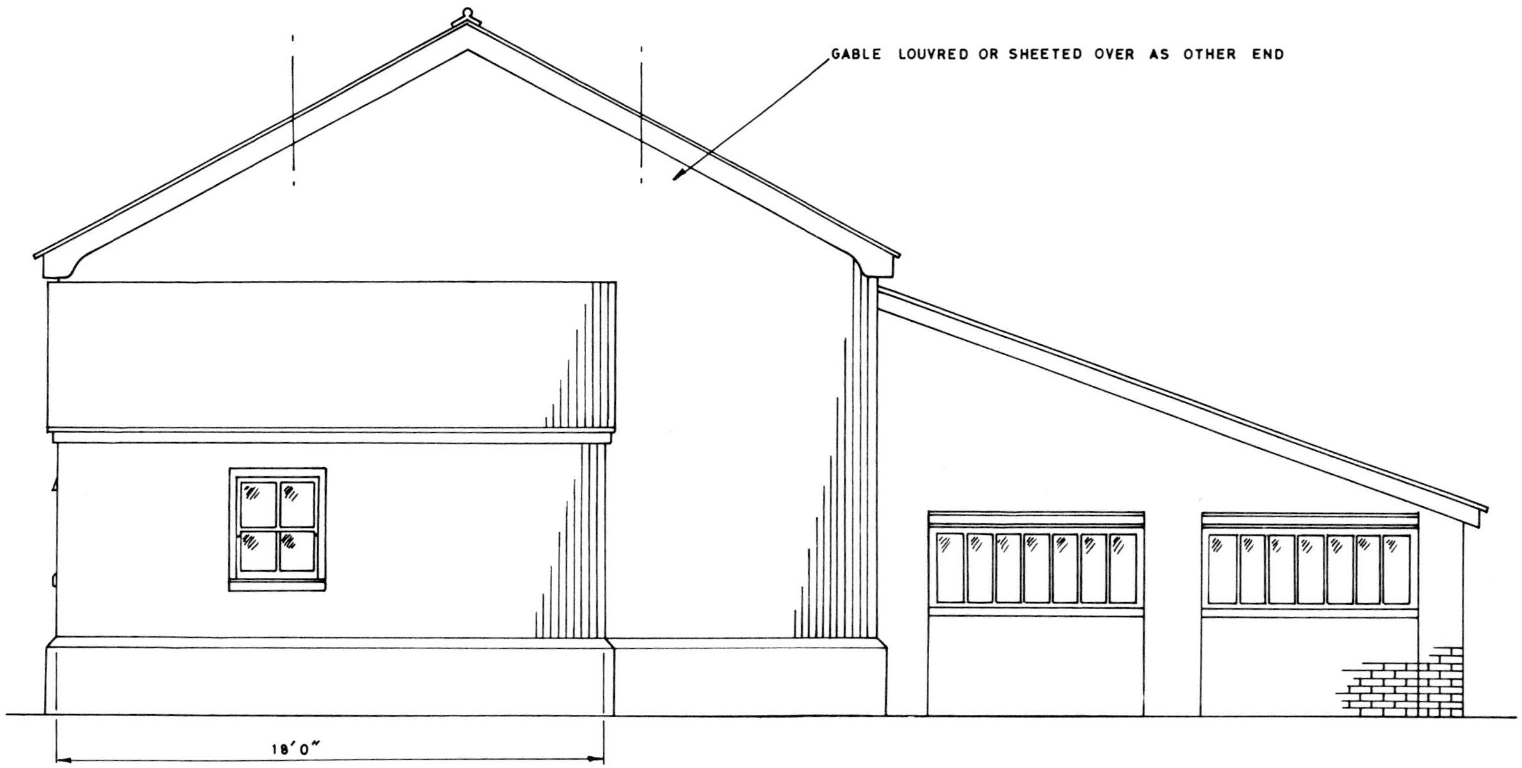

ROLVENDEN LOCO. SHED
K.&E.S.R.
(sheet 4)

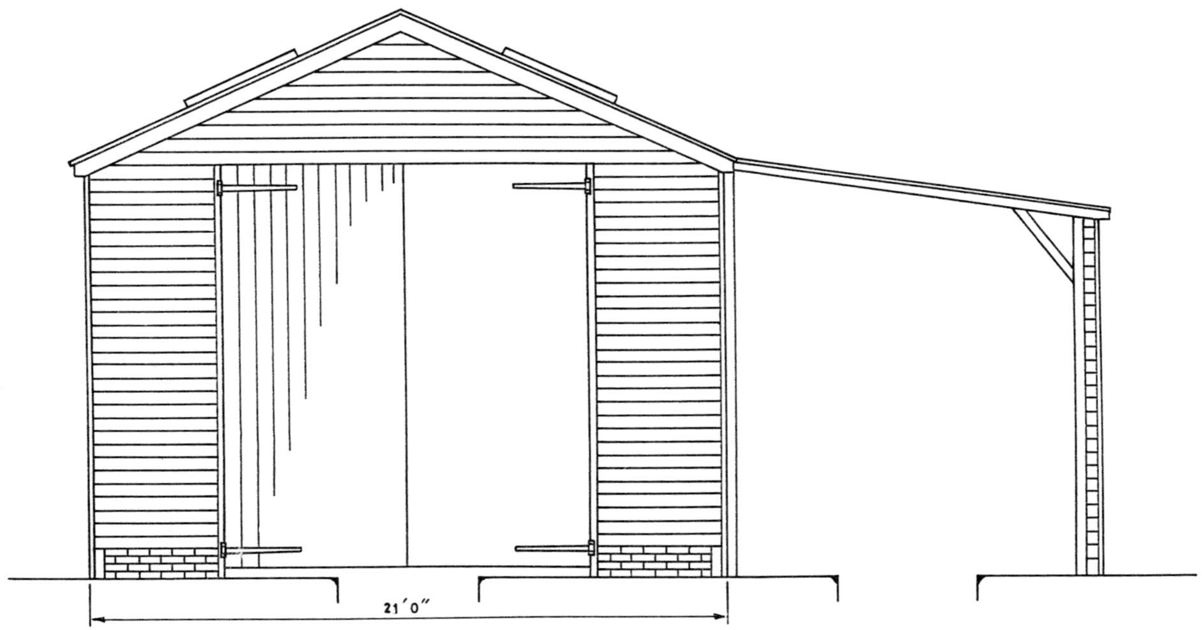

End ‘A’ — End ‘B’ is a mirror image

Scale :- feet

ROLVENDEN PAINT SHOP

K.&E.S.R.

(sheet 1)

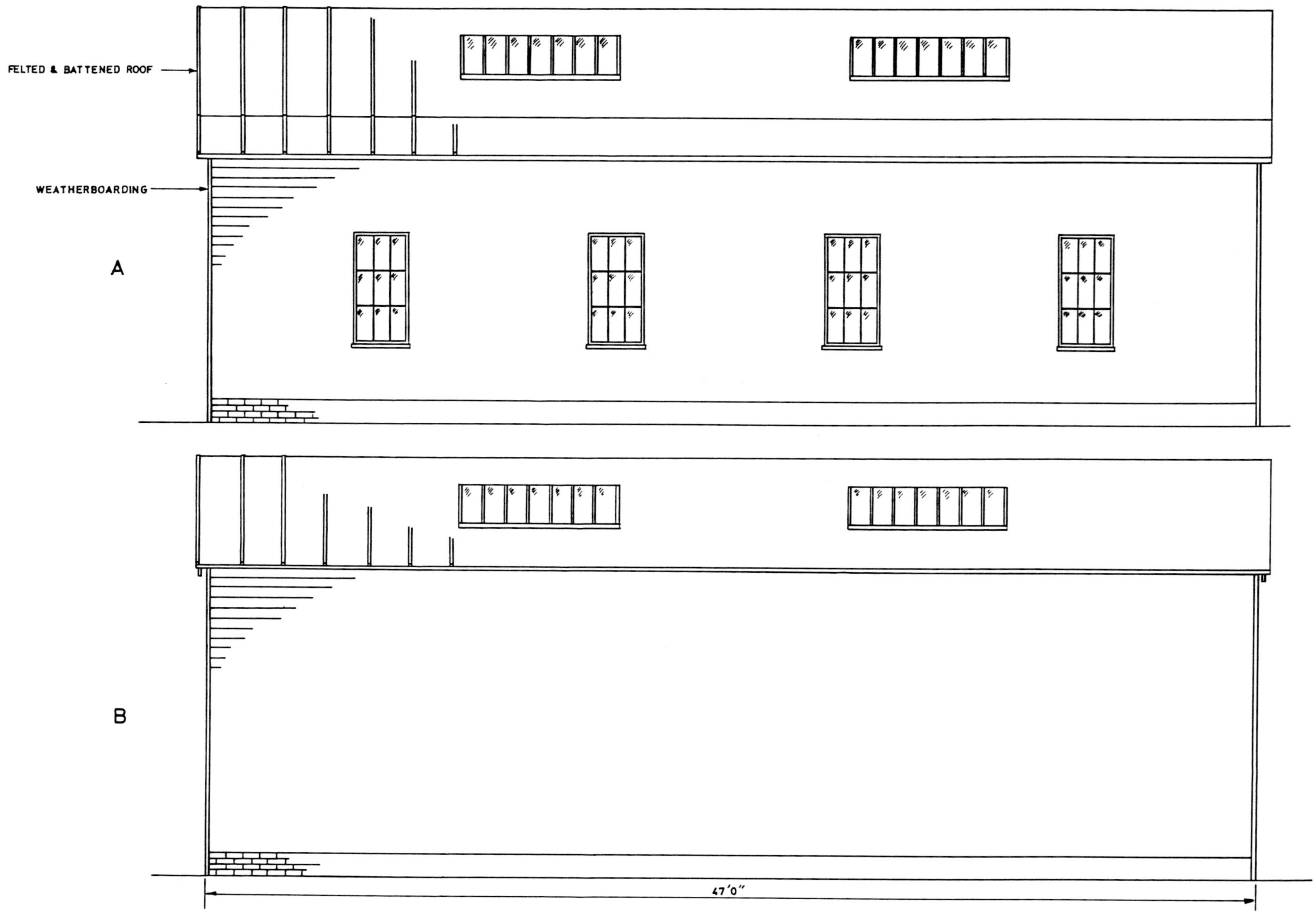

ROLVENDEN PAINT SHOP

K.&E.S.R.

(sheet 2)

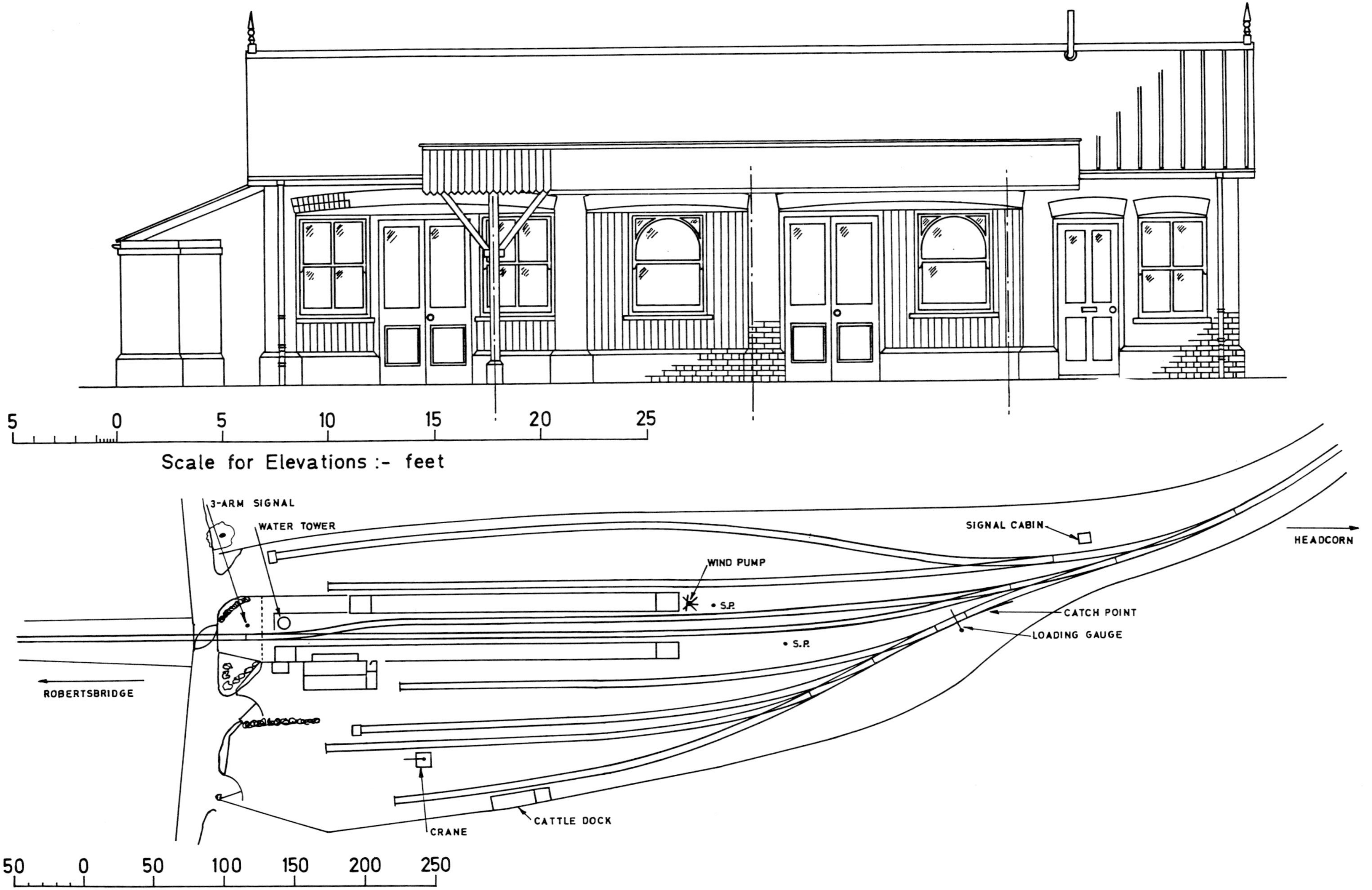

TENTERDEN TOWN STATION

K.&E.S.R.

(sheet 1)

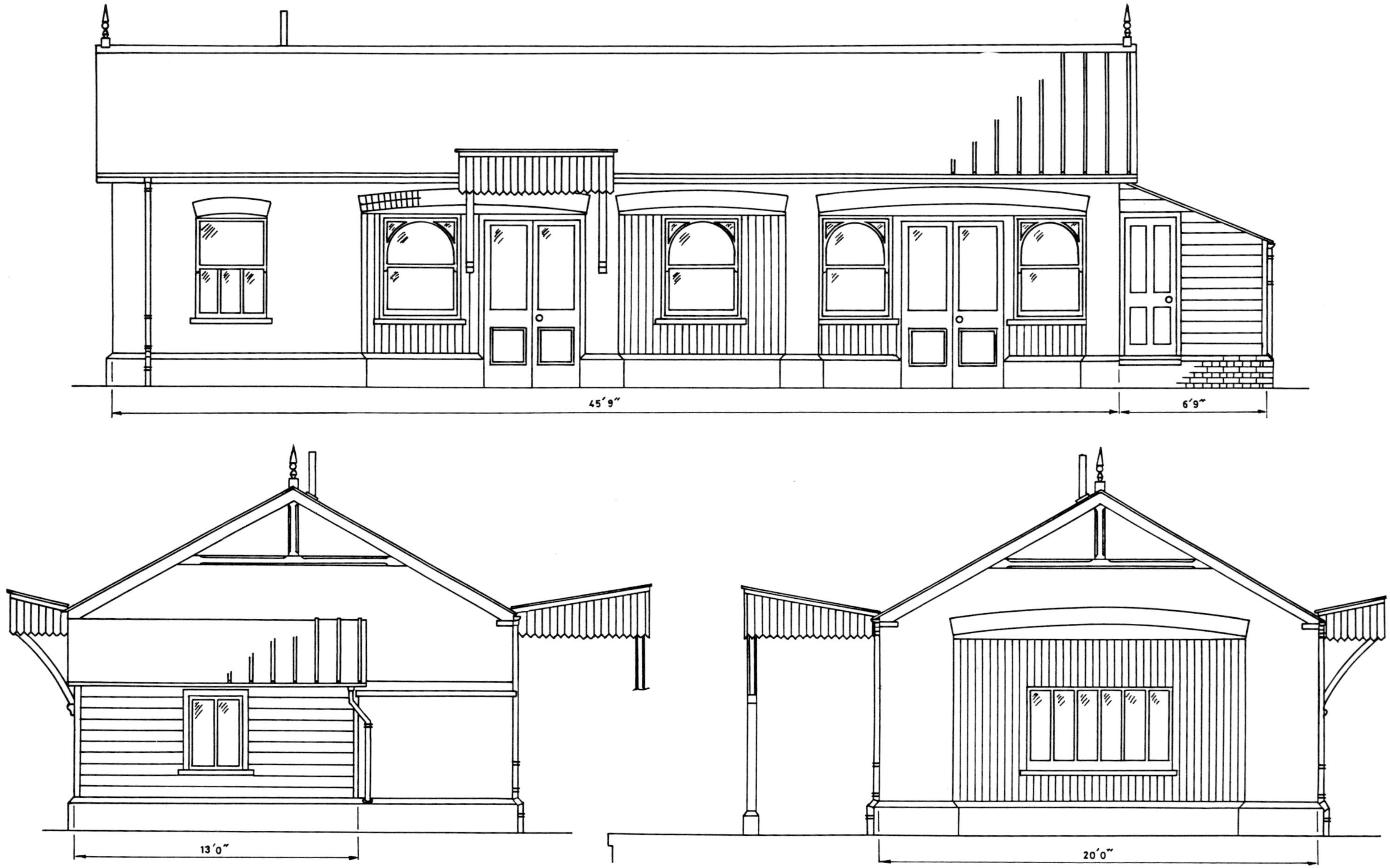

TENTERDEN TOWN STATION

K.&E.S.R.

(sheet 2)

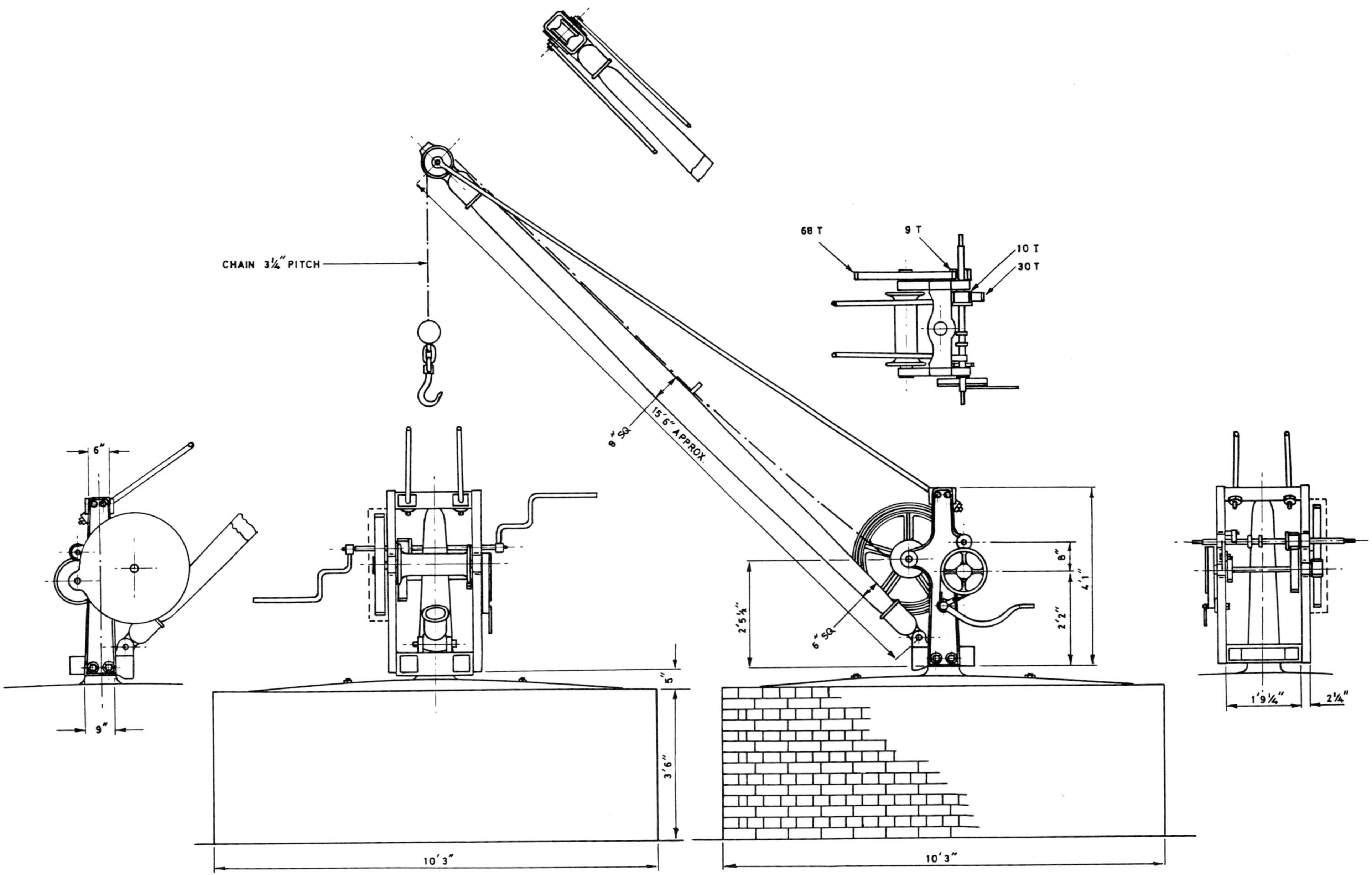

TENTERDEN TOWN YARD CRANE

K.&E.S.R.

Scale: 12 ins. 0 1 2 3 4 5 6 7 8 ft.

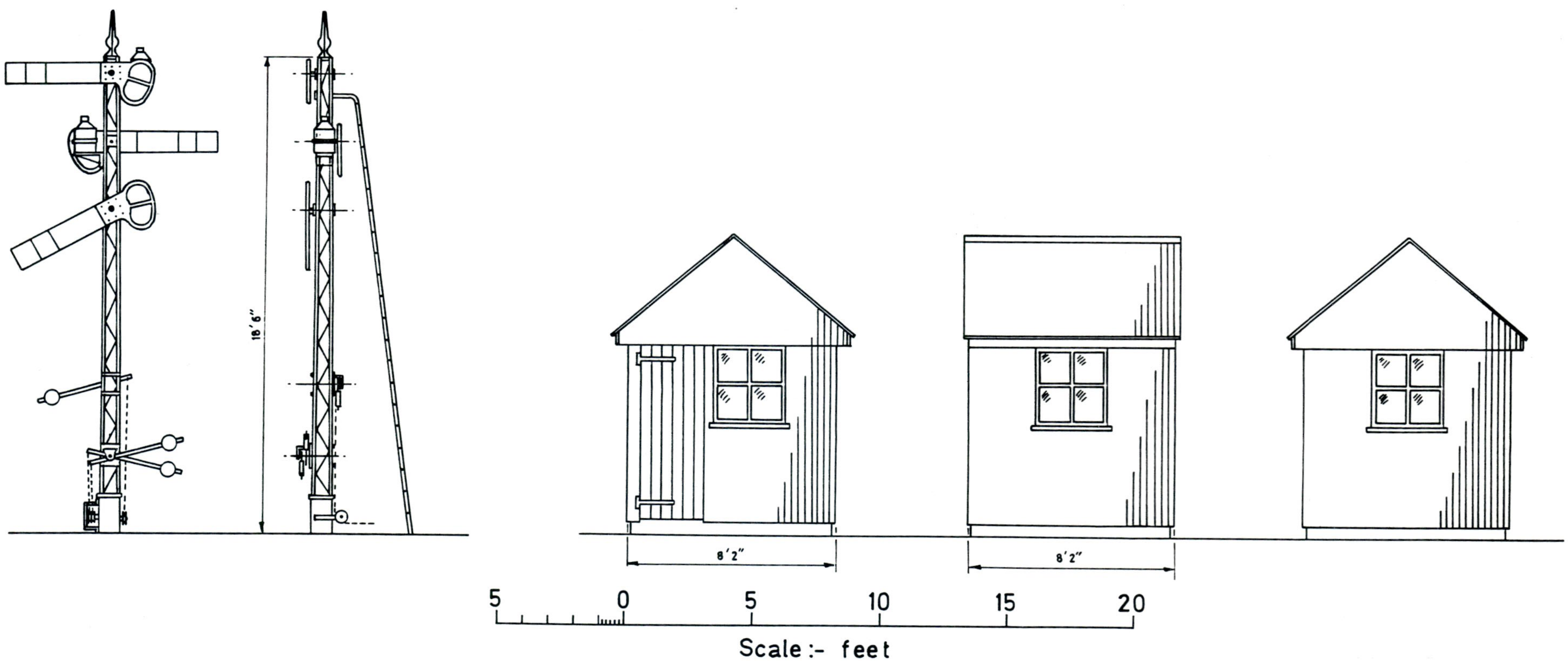

TENTERDEN TOWN

K.& E.S.R.

3-ARM SIGNAL & SIGNAL CABIN

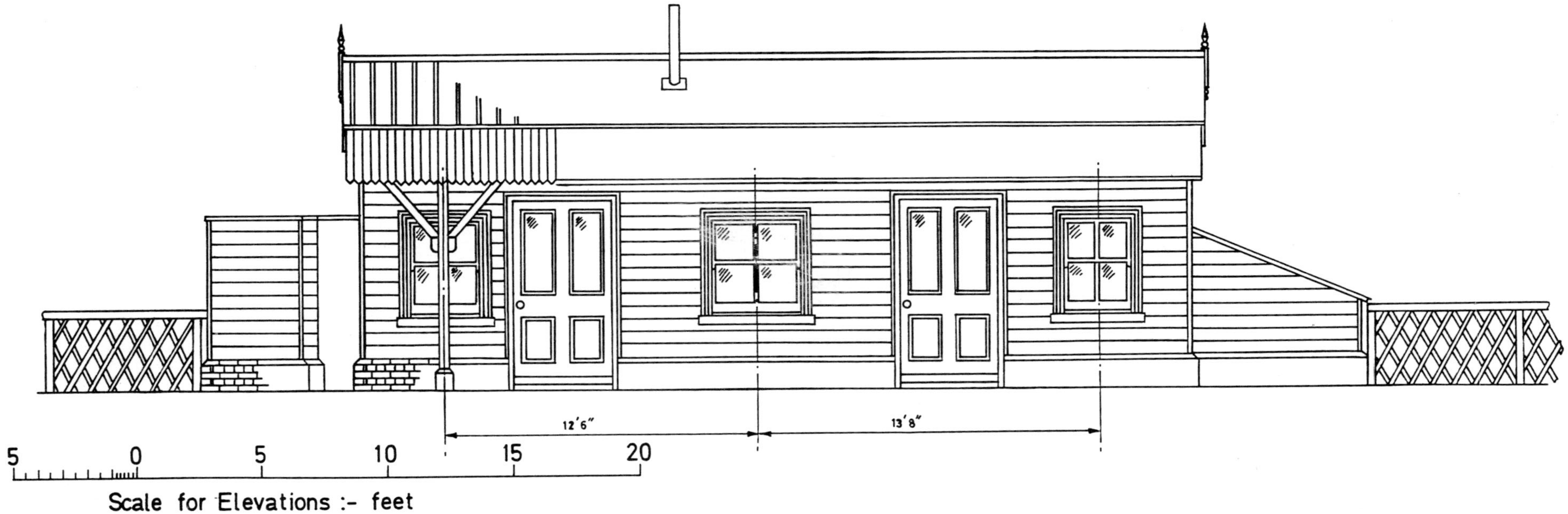

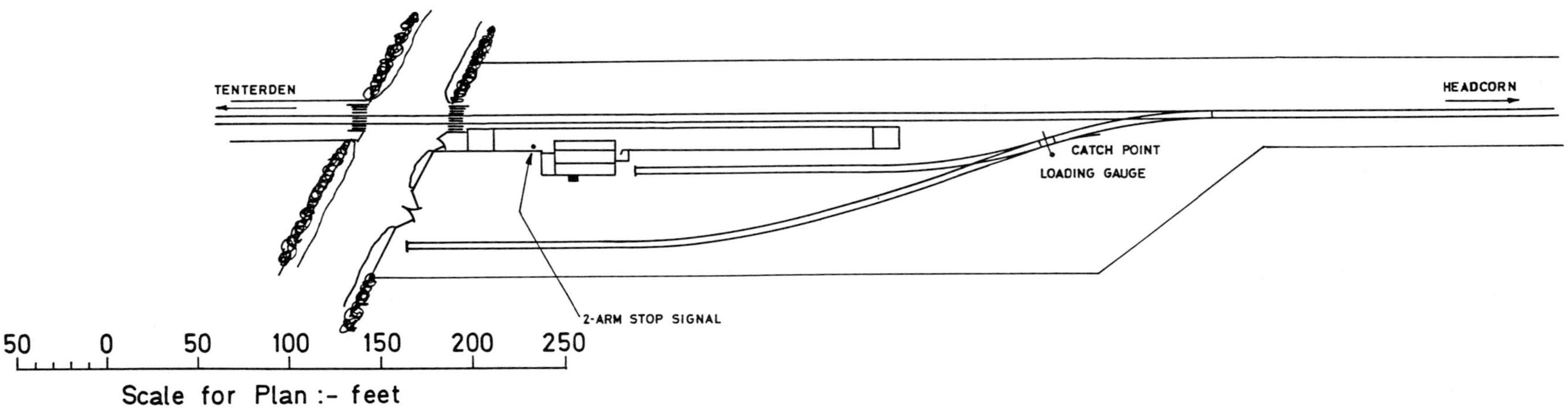

HIGH HALDEN ROAD STATION

K.&E.S.R.

(sheet 1)

HIGH HALDEN ROAD STATION

K.&E.S.R.

(sheet 2)